Jamal Koubachi

Molecular Spectroscopy Volume 1

Jamal Koubachi

Molecular Spectroscopy Volume 1

Ultraviolet/Visible (UV) Spectroscopy and infrared (IR) spectroscopy, Courses and corrected exercises

ScienciaScripts

Imprint
Any brand names and product names mentioned in this book are subject to trademark, brand or patent protection and are trademarks or registered trademarks of their respective holders. The use of brand names, product names, common names, trade names, product descriptions etc. even without a particular marking in this work is in no way to be construed to mean that such names may be regarded as unrestricted in respect of trademark and brand protection legislation and could thus be used by anyone.

Cover image: www.ingimage.com

This book is a translation from the original published under ISBN 978-620-6-71741-6.

Publisher:
Sciencia Scripts
is a trademark of
Dodo Books Indian Ocean Ltd. and OmniScriptum S.R.L publishing group

120 High Road, East Finchley, London, N2 9ED, United Kingdom
Str. Armeneasca 28/1, office 1, Chisinau MD-2012, Republic of Moldova, Europe
Printed at: see last page
ISBN: 978-620-3-49591-1

Molecular Spectroscopy Volume 1

Ultraviolet/visible (UV) spectroscopy

and infrared (IR) spectroscopy,

Courses and corrected exercises

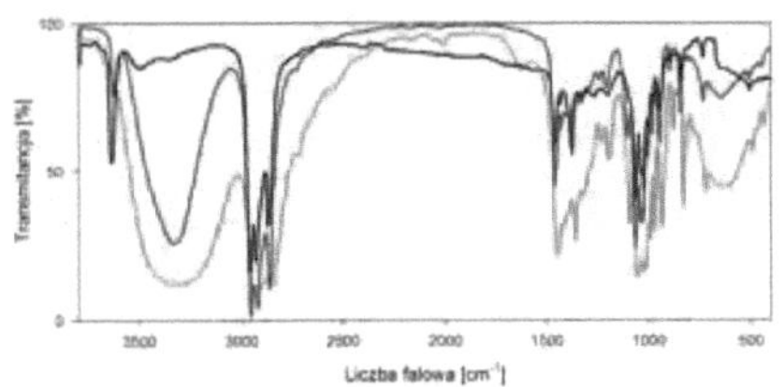

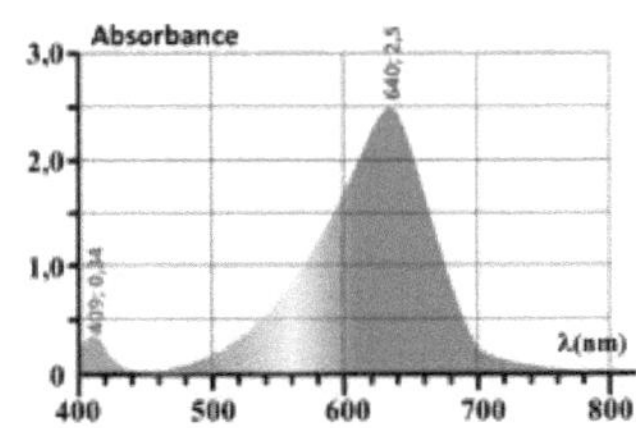

Prof. Koubachi Jamal
Taroudant Polytechnic Faculty
Ibn Zohr University, Morocco

Foreword

This book is designed for Master's students in analytical and quality control procedures, chemical analysis and organic chemistry. It can also be used by students studying for bachelor's degrees in pharmacy, chemistry, biology or agri-food.

This textbook with corrected exercises is designed to teach chemists the spectroscopic methods of analysis and identification techniques used to characterize synthetic organic or natural products by synergistically combining information from infrared and ultraviolet (UV) spectra. Basically, the molecule is perturbed by energetic probes and the responses are recorded in the form of spectra. These spectroscopic methods of analysis are often found in quality control laboratories.

The aim of this book is to refresh certain basic notions of analytical chemistry and to introduce students to physicochemical analysis techniques, so that they can cope with the difficulties that may be encountered when carrying out analyses and tests in the laboratory.

This book is presented in a simplified way, with several examples of spectra given to facilitate understanding. Application exercises are provided at the end to reinforce understanding and consolidate knowledge.

This document is divided into three chapters:

Chapter I: Basics of spectroscopy

Spectroscopy is the study of the interaction between electromagnetic radiation and matter. It provides information on the identity, structure and energy levels of atoms and molecules.

Chapter II: UV-Visible spectrometry.

In this chapter, we present the basics of UV-Visible spectrometry in a general way and go on to develop ultraviolet-visible molecular spectroscopy, references and corrected exercises. The aim is to interpret UV-Visible spectra of chromophores characterized by their maximum wavelength λ_{max} expressed in nm.

Chapter III: Infrared spectrometry

Chemists will be more at ease with infrared spectrometry if they have a minimum knowledge of theory and instrumentation. The characteristic absorptions of the groups are reviewed and pertinently associated with absorption diagrams of the groups, characteristic spectra, references and corrected exercises.

Table of contents

Introduction

Spectroscopy results from the interaction between matter and an electromagnetic wave. It has virtually replaced the former qualitative study of chemical compounds:

It enables structure to be determined on very small quantities of material, uses non-destructive methods and is extremely accurate.

Table 1 lists some of the characteristics of the different spectroscopies that chemists may need to use.

Table 1: *Different spectroscopic methods.*

	λ	v	**Radiation type**
NMR under a few teslas	0.1 to 100 m	3 to 3000 MHz	Radio waves and microwaves
Vibration-rotation	0.2 to 50 mm	6 to 1500 MHz	Infrared
Electronic transitions	> 10 mm	$< 3 \times 10^{16}$ MHz	Near UV, visible and near IR
Ionization	0.3 to 30 mm	10^{16} to 10^{18} MHz	X-rays (soft for valence levels, hard for deep levels)

Matter-radiation interaction occurs when a photon of energy $E = h\,v$ can be absorbed by a molecule. This phenomenon occurs on condition that there is a possible transition between two energy levels distant from E.

We will limit ourselves here to a simplified presentation of preliminary notions of spectroscopy, ultraviolet/visible (UV) spectroscopy and infrared (IR) spectroscopy as tools for analyzing and determining molecular structures.

Chapter I: Basics of spectroscopy

I. Introduction

1. Definition

Spectroscopy is the set of techniques used to analyze the light emitted by a light source and the light transmitted or reflected by an absorbing body. The interaction of light with matter is at the origin of most of the electrical, magnetic, optical and chemical phenomena observed in our immediate environment. Spectroscopy is also the study of electromagnetic radiation emitted, absorbed or scattered by atoms or molecules. It provides information on the identity, structure and energy levels of atoms and molecules by analyzing the interaction of electromagnetic radiation with matter.

2. Electromagnetic wave

It's an oscillation of electric and magnetic fields propagating through a medium. Electric field E and magnetic field H propagate perpendicularly and in phase (Figure 1).

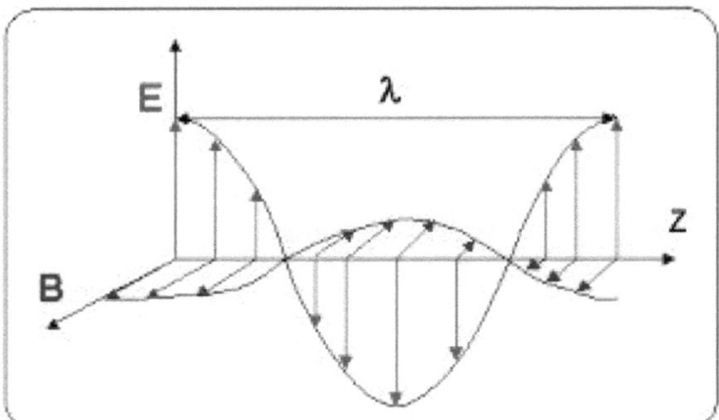

Figure 1: Electromagnetic wave

An electromagnetic wave is characterized by its frequency, wavelength or wavenumber.

Wavelength λ

Frequency $\square : \upsilon = \dfrac{c}{\lambda}$ (c: speed of light in a vacuum c = 3.10^8 m/s)

Period T : $T = \dfrac{1}{\upsilon}$

The electromagnetic energy of radiation is related to the above quantities by Planck's fundamental relation: $E = h\upsilon = \dfrac{hc}{\lambda}$ (h: Planck's constant h = $6.626.10^{-34}$ j.s).

3. Electromagnetic spectrum

It is made up of the continuous set of known electromagnetic radiations classified according to their frequency, wavelength, wave number or energy. It is shown in the figure below:

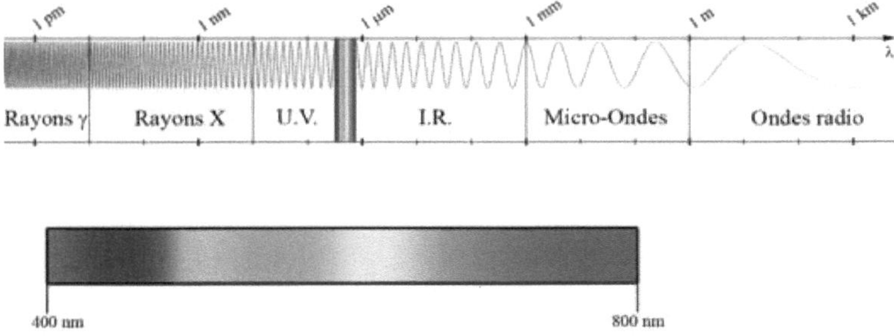

Figure 2: Electromagnetic spectrum

4. The corpuscular theory of electromagnetic radiation

Wave propagation is well described by wave theory. However, the interaction of radiation with matter has led us to attribute a corpuscular nature to electromagnetic waves.

- Corpuscular nature: The undulatory nature of light alone is not sufficient to interpret the interaction between light and matter.

Electromagnetic radiation behaves like a stream of particles, photons or quanta, moving at the speed of light.

Planck then Einstein proposed the quantum theory:

Light is made up of grains of energy: photons.

The photon is a particle that propagates at the speed of light and possesses a quantum of energy. The energy of a photon is given by Bohr's equation :

$E = h\nu$; where $h = 6.624.10^{-34}$ J.s is Planck's constant and ν is the wave's classical frequency.

5. Duality of matter

De Broglie's postulate

After Einstein's quantization of the electromagnetic field, giving it a corpuscular character, Louis de Broglie, in turn, took up the "onde-corpuscule" dualism and attributed the wave aspect to every particle making up matter:

Just as waves can behave like particles, particles can behave like waves.

In 1924, De Broglie associated a wavelength with any material particle endowed with a momentum p=mv, known as a De Broglie wave: $\lambda = h/p$ where h is Planck's constant.

II. The evolution of spectroscopy and fields of application

Isaac Newton is the founder of spectroscopy. In 1666, he was the first to understand that the spreading out of the seven colors of the rainbow by a prism is linked to the nature of light. These colors are in fact a succession of visible radiations of continuously varying wavelengths. The first spectroscope was built by Isaac Newton.

In 1800, William Herschel discovered the thermal effects of infrared radiation.

In 1803, Inglefield suggested that there might be invisible rays beyond the violet. The existence of these ultraviolet rays was demonstrated by Ritter and Wollaston

Spectroscopy really began with Bunsen and Kirchhoff (1824-1887). One of the first applications of spectroscopy was to test the chemical composition of the Sun and stars.

At the beginning of the 20th century, the development of equipment enabled spectroscopy to be widely used for a variety of applications.

Spectrophotometry is used in a variety of fields: chemistry, pharmaceuticals, environment, food processing, biology, etc., both in the laboratory and on industrial sites.

Example: In the pharmaceutical industry, many drug assays are performed using UV-VIS absorption spectrophotometry.

III. Principle of spectroscopy

Electromagnetic radiation absorbed by matter excites it, providing information capable of identifying and elucidating its structure.

Among the phenomena likely to be observed :

- Vibration and rotation of chemical bonds when the radiation absorbed is infrared.

- Jumps of electrons located in a valence level from one ground state to another when the absorbed radiation is ultraviolet-visible.

- Rotation of nuclear spins when the absorbed radiation is in the form of radiofrequency waves.

Whatever phenomenon is triggered, the molecule moves from a ground state characterized by an initial energy E_I to a final or excited state characterized by a final energy E_F (this is a less stable level where E_F is greater than E_I). During this excitation, there is a transition between the initial and final states.

If this transition occurs from the initial state to the final state, we say there's an absorption.

If the transition takes place in the opposite direction, we say there's an emission (Figure 3).

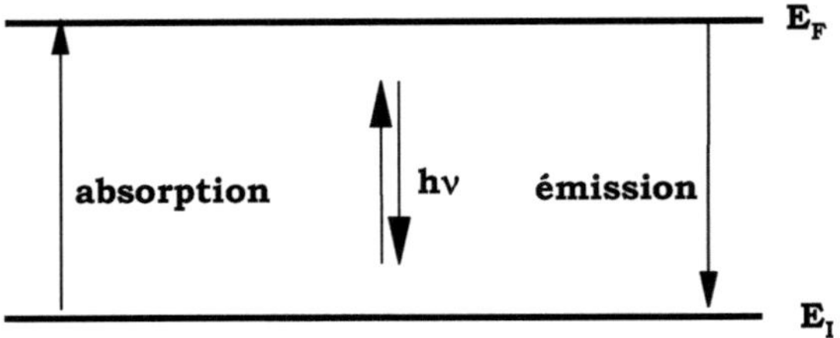

Figure 3: Absorption - Emission

There is an energy difference ΔE between the initial and final states

expressed by the relationship: $\Delta E = E_F - E_I = \eta \nu$

Please note:

An excited state is a very unstable state, with a very short lifetime (μs).

IV. Radiation-matter interaction

1. Molecular energy levels: energy states

The energy of atoms and molecules is quantified. This energy can come from a variety of sources:

- Molecules: rotation, vibration and electronic energy.

- Atoms: electronic energy.

There are 4 modes of movement, and therefore energy, for molecules:

⇨ Translation

⇨ Rotation

⇨ The vibration

⇨ Electronics (cloud deformation)

A first simplification consists in separating the uniform translational motion of the whole molecule, whose energy is not quantified.

Next, we distinguish between electrons and nuclei, particles with very different masses (nuclei are 10^3 to 10^5 times heavier). The movements of electrons are therefore "much faster" than those of nuclei.

Electron movements can be studied by considering nuclei as fixed (Born-Oppenheimer approximation).

This means separating the energies: on the one hand, the electronic energy E_e and, on the other, the energy due to the motion of the nuclei, of which there are two components: the vibration energy E_v and the rotation energy $E_{.r}$

An elementary particle (atom, ion or molecule) can only exist in certain quantified energy states. In the case of a molecule, total energy is taken to be the sum of the terms :

$$E = E_{Electronic} + E_{vibration} + E_{rotation} + E_{spin}$$

The orders of magnitude are very different: $E_e \gg E_v \gg E_r \gg E_{.s}$

The levels of electronic energy, vibration and rotation are represented by a diagram in which each level is represented by a horizontal line (Figure 4).

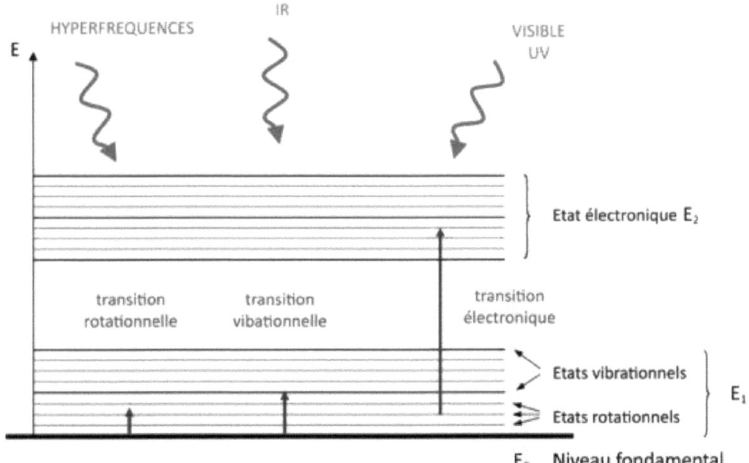

Figure 4: Diagram of electronic energy, vibration and rotation levels

2. Population distribution by state

Each elementary particle (atom, ion or molecule) has a unique set of energy states. The particle can be in any of these states.

The number of particles on a given energy level is called the population. The population on a level i relative to the population on the fundamental level obeys the Maxwell-Boltzmann distribution law (figure 5):

$$N_i / N_0 = (g_i / g_0)e^{-((E_i-E_0)/ kT)}$$

N_i : number of particles in the excited state i

N_0 : number of particles in ground state 0

g_i and g_0 : degeneration of states i and 0 respectively

E_i and E_0 : energy of states i and 0 respectively

k: Boltzmann constant $(1.38.10^{-23}$ J.K $)^{-1}$

T: temperature in Kelvin.

Figure 5: The Maxwell-Boltzmann distribution law

At room temperature, the thermal agitation, KT, is around 2.5 kJ/mol. The first excited vibrational level and the first excited electronic level have energies above this value (Figure 6).

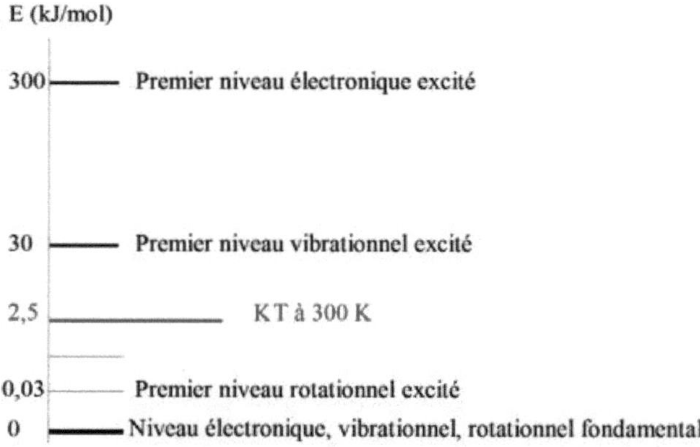

Figure 6: Energy diagram of the different levels

3. Wave-matter interaction: The different processes of radiation-matter interaction

Energy exchanges between matter and radiation can only take place in quanta: $\Delta E = h.\nu$

Four processes underlie spectroscopic phenomena: absorption, spontaneous emission, stimulated emission (in the case of lasers) and scattering.

The interaction of electromagnetic radiation with matter can take many different forms, and we will distinguish between the processes that underlie all spectroscopic phenomena: absorption, emission and scattering.

3.1. Absorption

When an atom is subjected to a light wave, it can absorb a photon. The atom, initially in a state of electronic energy E_a , then passes into an electronic state of higher energy $E_b > E_a$ (Figure 7).

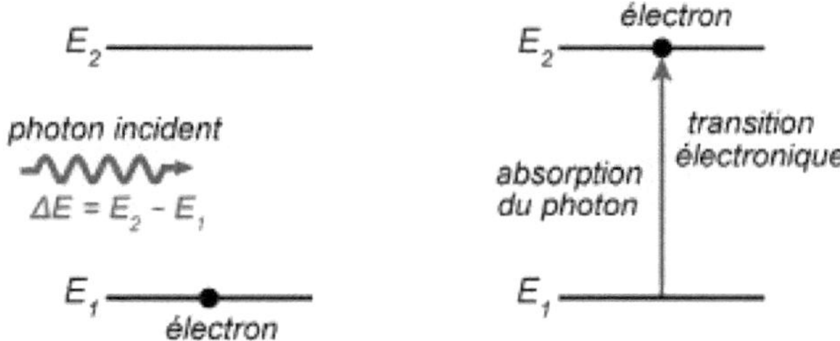

Figure 7: Absorption

3.2. Emission

The presence of incident radiation can induce an excited atom to emit a photon with the same characteristics as the incident photons.
This process is the basis of laser operation (figure 8).

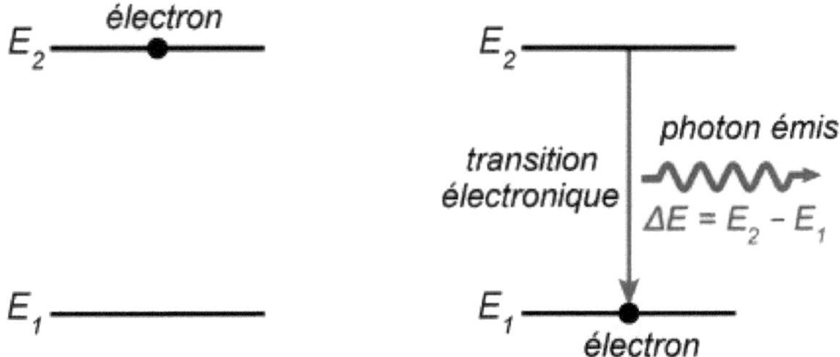

Figure 8: Emission

3.3. Diffusion

Scattering is the phenomenon whereby radiation, such as light, sound or a moving particle, is deflected in multiple directions by interaction with other objects (Figure 9).

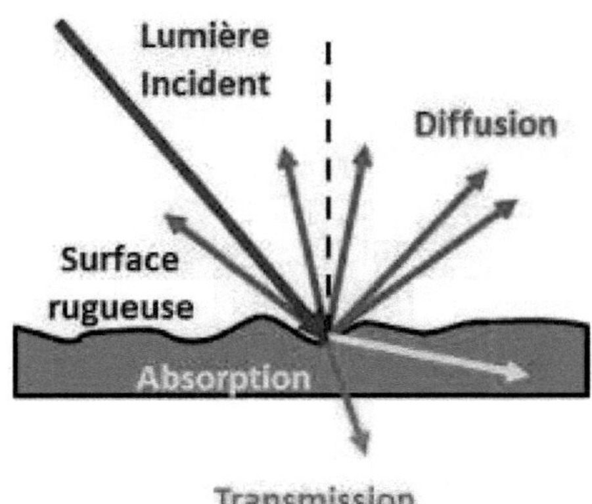

Figure 9: Diffusion

As a result of the energy exchange, electromagnetic radiation causes a disturbance in internal molecular motion. A transition occurs from one energy level to another, depending on the induced motion.

Table 2: Effect of electromagnetic radiation on matter

Absorbed radiation	Effect on material
Radio waves	Nuclear spin transitions (NMR)
Microwave	Molecular rotation. Electron spin transitions (electron paramagnetic resonance EPR)
Infrared	Molecular rotation and vibration
Visible and ultraviolet	Valence electron skipping
X-rays	Extraction of electrons from the atom's inner layers

V. Selection rules

Since wave-matter interaction is a quantum phenomenon, it is accompanied by selection rules.

Selection rules determine whether a transition is permitted or prohibited. The wave-molecule interaction can only take place if :

- the light frequency corresponds to the energy difference (Δ E) between the levels concerned

- the motion causes, at the same frequency, the variation of the dipole moment μ of the system.

If μ is the electric dipole moment, then the transitions are of the electric dipole type (responsible for the phenomena observed in the UV, visible and IR).

If μ is the magnetic dipole moment, the transitions are of the magnetic dipole type (responsible for nuclear magnetic resonance and electron paramagnetic resonance phenomena).

VI. Spectrum presentation

This is a two-dimensional diagram (Figure 10):

Abscissa: We carry

- or the wavelengthλ in cm for the microwave range, inμ m for the IR range and in nm for the UV-visible range.

- or the wave number in cm^{-1} whatever the domain concerned.

The ordinates:

Two quantities can be used: transmission and absorbance.

Spectra presentation:

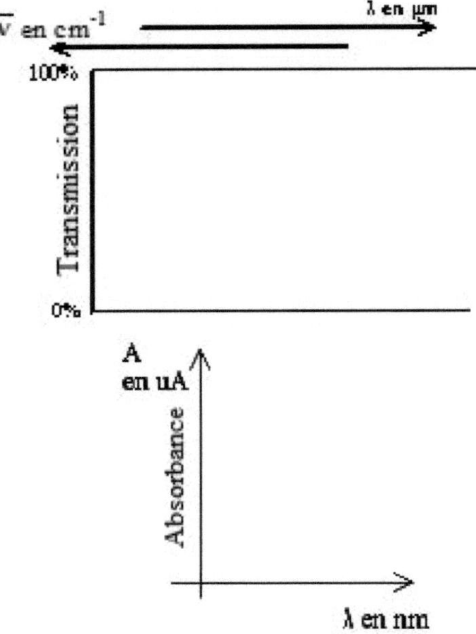

Figure 10: IR and UV-visible spectra

1. Line spectra

In an atom, a variation in electronic energy gives rise to a single spectral line. The position of each line corresponds to monochromatic radiation (Figure 11).

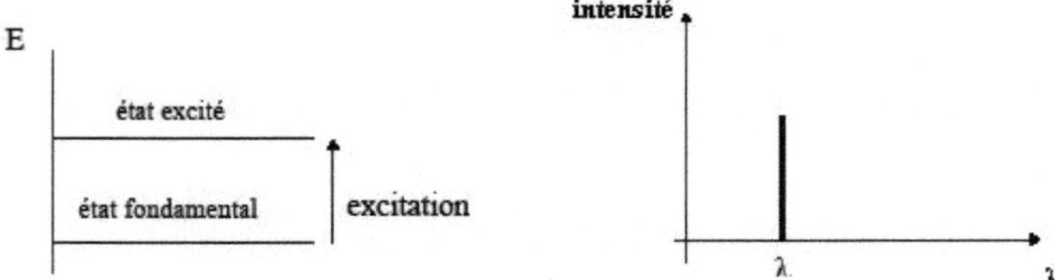

Figure 11: Line spectrum

2. Band spectra

Theoretically, the spectrum of a molecule is a line spectrum (quantization, discrete energy values). However, experimentally, for example, a transition between two electronic levels can lead to a modification of both vibrational and rotational energies, and thus to a set of transitions of very similar energies, resulting in a band spectrum (Figure 12).

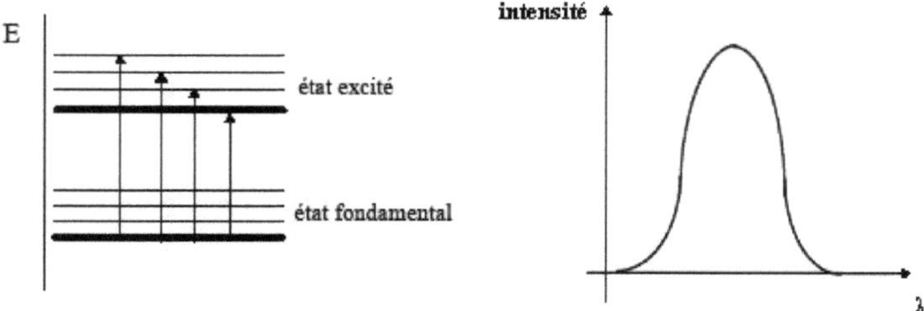

Figure 12: Band spectrum

VII. Beer-Lambert law

1. Concept

When a beam of electromagnetic radiation (IR or UV-visible) passes through a cylindrical vessel of width "l" containing a compound in solution, the intensity of the incident radiation I_0 decreases if the compound absorbs a certain amount of the radiation I_A (Figure 13).

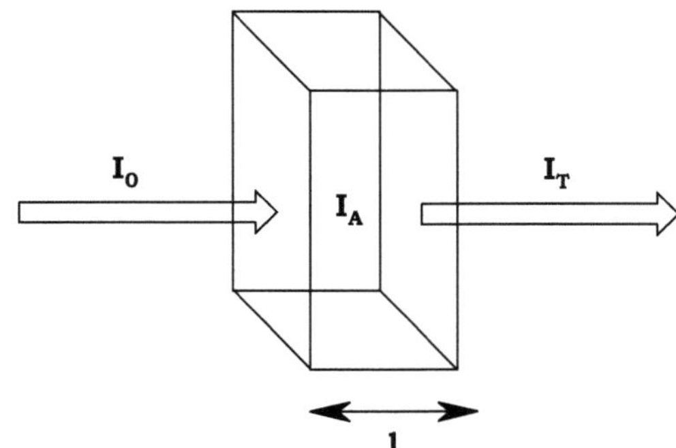

Figure 13: Concept of Beer Lambert's law

We give:

I_0 : intensity of incident electromagnetic radiation

I_T : intensity of transmitted electromagnetic radiation

I_A : amount of electromagnetic radiation absorbed

2. Transmittance

Transmittance T is the percentage of electromagnetic radiation transmitted by a solution. It is defined by the relationship (Figure 14):

$$T\% = \frac{I_T}{I_0} x100$$

Figure 14: Transmittance T

Note: Transmittance is often used in infrared applications.

3. Absorbance

Reflects the amount of electromagnetic radiation absorbed. It is expressed by the following relationship (Figure 15):

$$A = \log\frac{100}{T}$$

Figure 15: Absorbance A

Note: Absorbance is often used in the UV-visible range.

4. Statement of the Beer-Lambert law

The absorbance of a compound in solution is directly proportional to concentration and cell thickness. It is also defined by the relationship (Figure 16):

$$A = \varepsilon\, l\, C$$

A: Absorbance of dimensionless solution

C: Concentration of compound in solution

l: Thickness or width of the tank, always expressed in cm, also called optical path.

ε: Molar extinction coefficient expressed in (L/mol.cm)

Figure 16: Beer Lambert's law

Please note:

Two different compounds **A** and **B** with the same concentration do not give the same absorbance. This is due to the difference in their molar extinction coefficients. This coefficient reflects a compound's ability to absorb electromagnetic radiation. It is characteristic of each compound.

Chapter II: Ultraviolet-visible molecular spectroscopy

I. Introduction

Spectroscopic techniques are based on the exchange of energy between radiant energy and matter. UV-visible spectroscopy is a method for analyzing and identifying chemical species.

UV-visible spectroscopy was one of the first methods used to probe matter [structure of molecules (qualitative), composition of solutions (quantitative)] using waves. We use the interaction between visible-UV light and matter, and more specifically the electrons in molecular orbitals (π-electrons and free pairs), which causes electronic transitions. Electrons change electronic layers in molecules.

In particular, absorption spectrophotometry focuses on the absorption of light radiation in the visible and near-ultraviolet regions of the electromagnetic spectrum (Figure 36).

Ultraviolet-visible (UV-visible) electromagnetic radiation is classified according to wavelength values:

Near-ultraviolet range: 200 nm $\leq \lambda$ δ 400 nm

Visible range: 400 nm $\leq \lambda$ δ 800 nm

The far UV (10 - 200 nm) is also concerned, although in this case work is carried out in a vacuum or inert gas atmosphere, as atmospheric oxygen covers the signals of other substances (Figure 17).

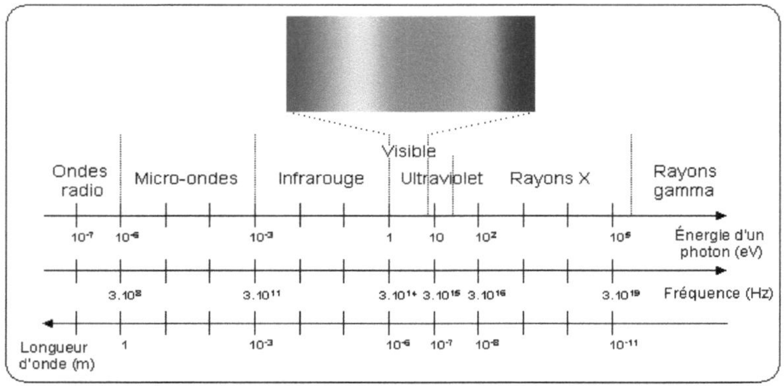

Figure 17: Electromagnetic spectra

The absorption of these types of radiation by molecules is capable of producing energetic transitions of the molecule's outer electrons, whether or not they are involved in a bond.

To observe an electron transition in the ultraviolet, or Visible, range, the molecule must possess electrons that are easily excited by this radiation. This is most often the case for unsaturated organic molecules (with chromophore groups) or inorganic species with d-layer electrons.

Remarks:

- UV-Visible cannot determine the total structure of a molecule.

- UV-visible wavelengths are shorter than IR wavelengths. As a result, UV-visible radiation is more energetic than IR.

- Color is the manifestation of visible light absorption (λ from 400 to 800 nm, i.e. from 0.4 to 0.8 μm).

- Substances that absorb only: in the near UV (ultraviolet) ($\lambda = 200$ to 400 nm) and IR (infrared: $\lambda > 800$ nm) appear colorless.

All substances absorb in the far UV ($\lambda < 200$ nm).

II. Selection principles and rules

A UV-Visible transition (often 180 to 750 nm) corresponds to a jump of an electron from an occupied fundamental molecular orbital to a vacant excited molecular orbital (Figure 18).

Matter then absorbs a photon whose energy corresponds to the energy difference between the fundamental and excited levels. But not all energetically possible transitions are possible.

Permitted transitions are those that cause a variation in the electric dipole moment.

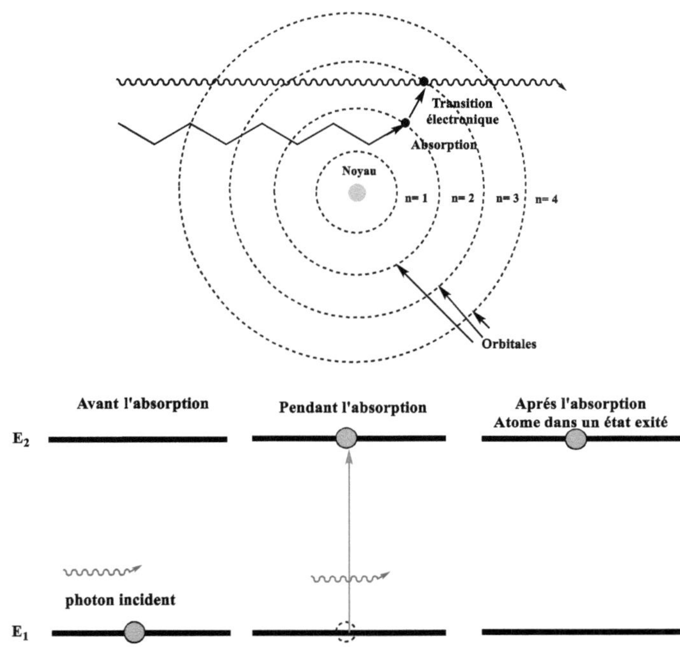

Figure 18: Electronic transition

III. Different types of electronic transitions and UV-Visible absorption

1. Absorption spectrum as a function of wavelength

By passing UV or visible radiation through a substance, we can obtain a residual intensity (absorption) curve as a function of wavelength. The result is an electronic absorption spectrum.

A UV-Visible spectrum is a plot of absorbance as a function of wavelength (in nm). The absorption band is characterized by its wavelength position (λ_{max}) and by its intensity related to the molar extinction coefficient ε_{max} ($A = \varepsilon\,\ell\,C$); the value of ε can indicate whether the transition is allowed or forbidden (Figure 19).

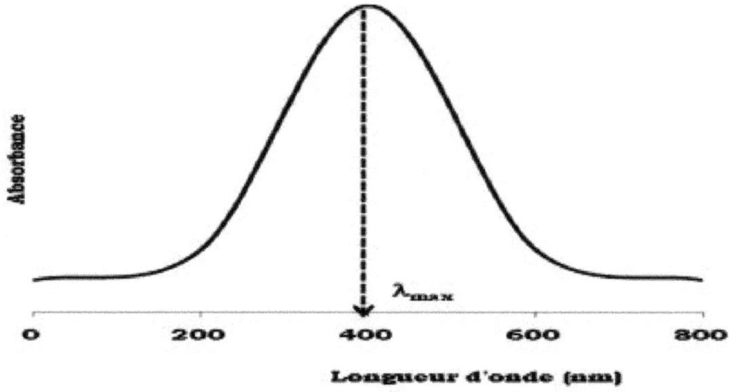

Figure 19: UV - Visible adsorption spectrum

Note:

A compound that absorbs in the UV-Visible range is characterized by its maximum wavelength λ_{max} expressed in nm.

For example:

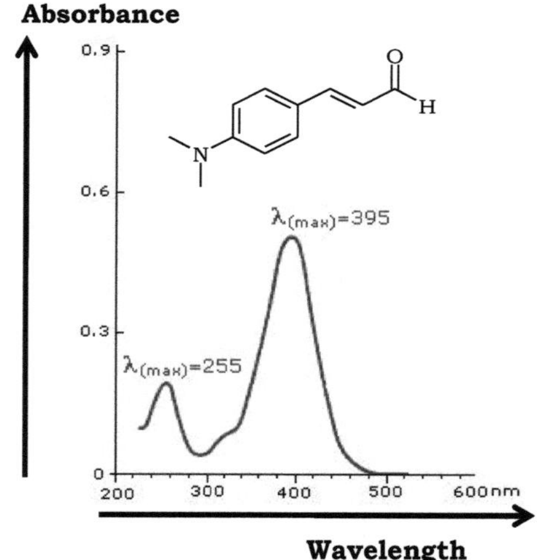

Figure 20: UV-Visible adsorption spectrum of (E)-3-(4-(dimethylamino)phenyl)acrylaldehyde

2. Electronic transitions

Electronic transitions are made between the different molecular orbitals.

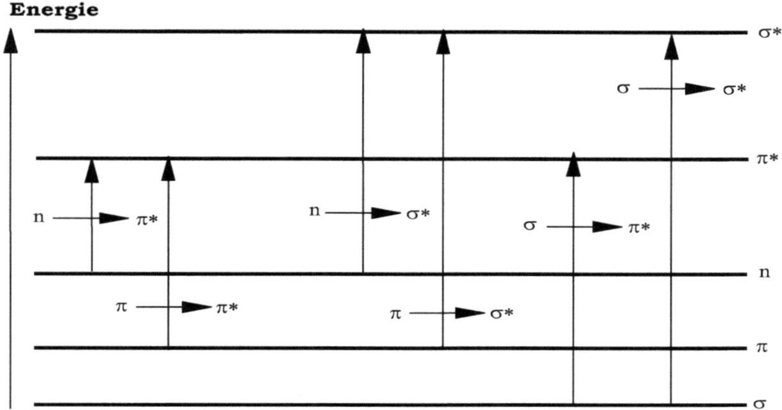

Figure 21: Electronic transitions between different molecular orbitals.

Each electronic state has a corresponding energy E_σ , E_π , E_n , $E_{\pi*}$ **and** $E_{\sigma*}$ as shown in the figure above. The electronic transition corresponds to an energy variation such as:

$$\Delta E = h\,\nu = \eta\ \chi/\lambda$$

Where λ corresponds to the wavelength of the electronic transition under consideration.

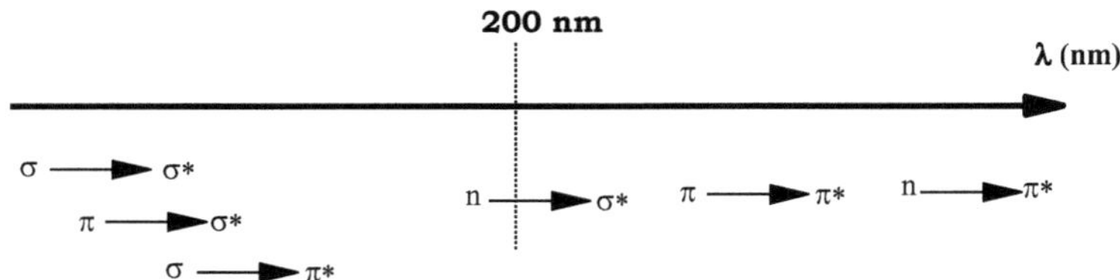

Figure 22: Wavelengths of different electronic transitions in the UV-Visible range

Note:

To observe an electronic transition in UV-Visible light, the molecule must possess unsaturation.

♣ Transition σ → σ*

It corresponds to the transition of an electron from a σ-bonding molecular orbital (MO) to a σ*-anti-bonding MO, which requires a lot of energy. Spectral range: far UV.

Example: hexane *(in the gaseous state)* C H , λ_{614max} = 135 nm, high-intensity transition.

Saturated hydrocarbons have only bonds of this type, and are transparent in the near-UV range.

Cyclohexane and heptane are used as near-UV solvents. At 200 nm, the absorbance *A* of a 1 cm thickness of heptane is equal to 1. Unfortunately, the solvating power of these solvents is insufficient to dissolve many polar compounds.

Similarly, water's transparency in the near UV (*A* = 0.01 for ℓ = 1 cm, at λ = 190 nm) is due to the fact that there can only be $\sigma \rightarrow \sigma^*$ and $n \rightarrow \sigma^*$ transitions.

♣ Transition n → σ*

It corresponds to the passage of an electron from a free doublet *n* of the O, N, S, Cl atoms to an anti-bonding OM σ^* leading to a transition of average intensity around 180 nm for alcohols, around 190 nm for ethers or halogenated derivatives and around 220 nm for amines.

The energies involved are generally lower than those of $\sigma \rightarrow \sigma^*$ transitions.

They correspond to wavelengths between 150 and 250 nm. The absorption coefficient varies from 100 to 5000 L.mol^{-1} .cm .$^{-1}$

This transition occurs at around 180 nm for alcohols, 190 nm for ethers and 220 nm for amines.

Example: Ethylamine λ_{max} = 210 nm (ε = 800); ether λ_{max} = 190 nm (ε = 2,000);

Methanol: λ_{max} = 183 nm (ε = 50); 1-chlorobutane: λ_{max} = 179 nm.

♣ Transition n → π*

It corresponds to the passage of an electron from a doublet of a non-bonding *n-type* OM to an anti-bonding π^* OM. It occurs in the case of a heteroatom doublet in an unsaturated system.

Example: the carbonyl group C=O (270 - 280 nm) has a low molar absorption coefficient of between 10 and 100 L.mol^{-1} .cm . $^{-1}$

Ethanal λ = 293 nm (ε = 12, in ethanol as solvent).

♣ Transition $\pi \rightarrow \pi^*$

Corresponds to ethylenic compounds. An electron passes from an OM π to an OM π^*, and it has a strong absorption band around 170 nm with an absorption coefficient ranging from 1000 to 10000 L.mol^{-1} .cm .$^{-1}$

Example: ethylene λ_{max} = 165 nm

The four types of transitions are grouped together on a single energy diagram to situate them in relation to each other in the general case, and to specify the spectral ranges involved (Figure 23).

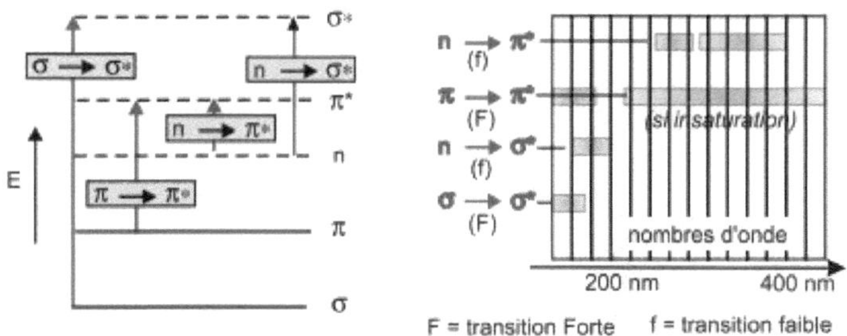

Figure 23: Comparison of the most frequently encountered transitions in simple organic compounds.

♣ Transition $d \rightarrow d$

It concerns transition metals with electrons in d-OMs. These complexes (or inorganic salts) are colored, and absorption in the visible range is often due to a d-d transition in which an electron passes from an occupied d orbital to a vacant, higher-energy d orbital.

For example, solutions of the metal salts titanium (Ti(H$_2$ O)]$_6^{+++}$ or copper [Cu(H$_2$ O)]$_6^{++}$ are blue, while potassium permanganate gives purple solutions, etc.

Absorption coefficients are generally very low, from 1 to 100 L.mol^{-1} .cm^{-1}

Table 3: Colors with wavelength λ

Color	λ (nm)	ν (10^4 Hz)	E (eV)	E (Kj/mol)
Red	700	4,28	1,77	171
Orange	620	4,84	2,00	193
Yellow	580	5,17	2,14	206
Green	530	5,66	2,34	226
Blue	470	6,38	2,64	254
Violet	420	7,14	4,15	285
Ultraviolet	< 300	> 15,0	> 5,00	> 400

3. UV-Visible absorbing compounds

3.1 Molecular electronic structure of dihydrogen

The simple covalent bond noted σ between the two hydrogen atoms results from an axial overlap of the two atomic orbitals (AO) $1s_A$ and $1s_B$ of the two hydrogen atoms to form molecular orbitals (binding OM σ) and (anti-binding OM σ* which is a virtual OM) (Figure 24).

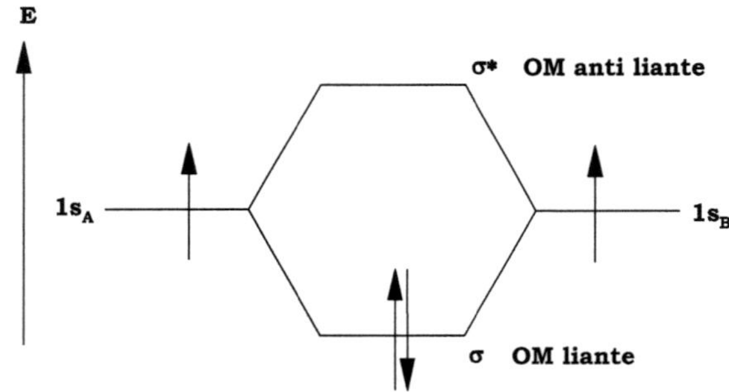

Figure 24: OM energy diagram for dihydrogen

Following external excitation, the molecular system undergoes an electronic transition from the most stable state (OM □) to the excited state (OM σ*) (Figure 25).

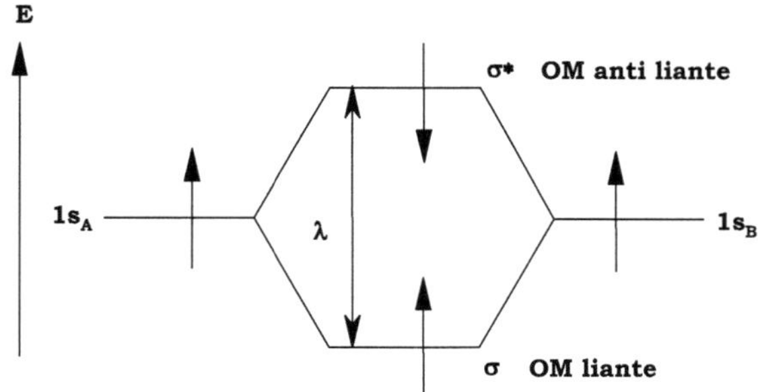

Figure 25: Electronic transition between σ ετ σ*

Electronic transitions can be observed between an occupied electronic level and an electronic level that can accept an electron.

For the dihydrogen molecule, the electronic transition has a wavelength □□□□□□□nm. However, this wavelength does not belong to the UV-Visible range. We shall see later that this electronic transition is not permitted in the UV-Visible range.

Note:

For an electronic transition to be observed in the UV-Visible range, the occupied OM must have a higher energy and the vacant one a lower energy. This is the case for compounds with π bonds.

3.2. Molecular electronic structure of ethylene

Consider the ethylene molecule with chemical structure: $CH_2=CH_2$.

The two carbon atoms are hybridized sp^2. It's a flat molecule. Each carbon has three bonds σ and one bond □. The π bond results from a lateral overlap of the 2p OAs of each carbon. Two cases are possible:

Both electrons simultaneously benefit from electrostatic interaction with the two carbon nuclei. The OM π is the result of the in-phase combination of two OAs of 2p(C).

The OM can be written: $\pi \cong 2p(C1) + 2p(C2)$

Both electrons can occupy this OM if they have opposite spins.

The two electrons cannot simultaneously benefit from the attraction of the two nuclei of the two C's. The OM π^* is the result of the reverse-phase combination of the two OA's of 2p(C).

The OM can be written: $\pi^* \cong 2p(C_1) - 2p(C)_2$

In this orbital, the electrons avoid being between the 2 C atoms (a virtual state).

The qualitative energy diagram for ethylene is shown as follows:

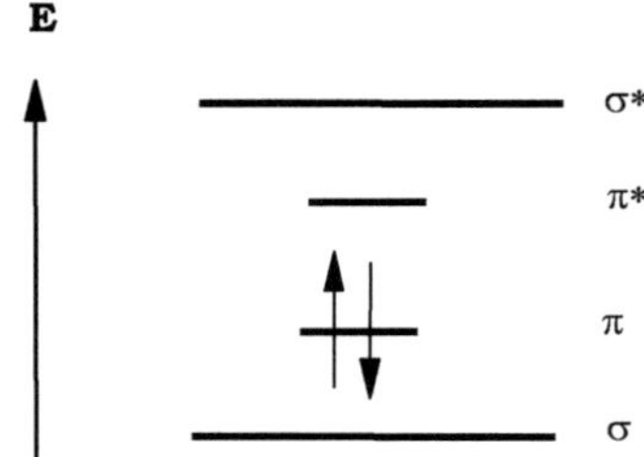

Figure 26: Qualitative energy diagram for ethylene

The electronic transition of ethylene is:

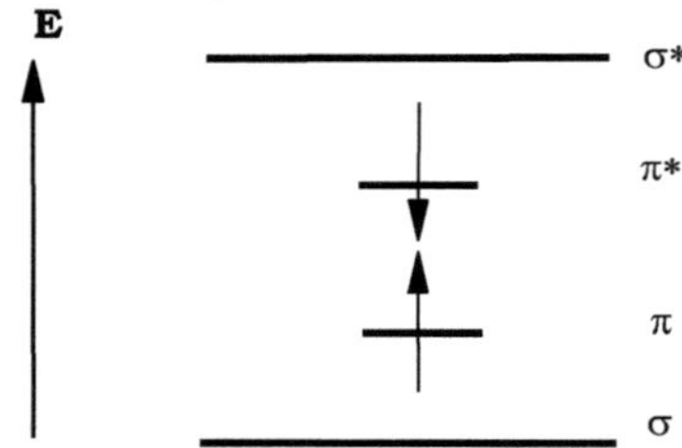

Figure 27: Electronic transition of ethylene

Ethylene absorbs at a wavelength oφ λ= 165 nm, which is close to the UV-Visible range.

The most easily excitable electrons in a compound are the π electrons of double bonds and the n electrons (free electron doublets on the peripheral layer of heteroatoms: N, O, S). It is these functional groups (also known as chromophore groups) that are responsible for a compound's absorption of UV-Visible radiation.

Table 4: Examples of electronic transitions for some compounds.

	Transition	max λ(nm)	ε (L.mol^{-1} .cm)$^{-1}$
Ethylene (C=C)	$\pi \rightarrow \pi^*$	165	15000
1-hexyne (C≡C)	$\pi \rightarrow \pi^*$	180	10000
Ethanal C=O)	$n \rightarrow \pi^*$	293	12
	$\pi \rightarrow \pi^*$	180	10000
Nitromethane (NO)$_2$	$n \rightarrow \pi^*$	275	17
	$\pi \rightarrow \pi^*$	200	5000
Bromomethane (CH$_3$ - Br)	$n \rightarrow \sigma^*$	205	200

4. Chromophores and definitions

A chromophore is a function or group of atoms that modify the frequency of the UV wave and the intensity of absorption ($\square$).

The functional groups of organic compounds (ketones, amines, nitro derivatives, etc.) responsible for UV/VIS absorption are called *chromophore groups* (Table nn). A species with a carbon skeleton that is transparent in the near-UV range and carries one or more chromophores is a chromogen.

Table 5: Characteristic chromophores for some nitrogen groups.

Name	Chromophore	max $\square$(nm)	ε_{max} (L.mol^{-1} .cm)$^{-1}$
Amine	-NH$_2$	195	3000
Oxime	= NOH	190	5000
Nitro	- NO$_2$	210	3000
Nitrite	- ONO	230	1500
Nitrate	- ONO$_2$	270	12
Nitroso	-N = O	300	100

Auxochrome grouping:

It is a saturated group linked to a chromophore and capable of modifying the wavelength and intensity of absorption (OH, NH$_2$, Cl,...).

The four terms used are:

🜄 Bathchrome effect:

Chromophore decreases absorption frequency (increases λ)$_{max}$

🜄 Hypsochrome effect:

Chromophore increases absorption frequency (decreases λ) $_{max}$

🜄 Hypochromic effect:

Chromophore reduces absorption intensity (decreases □)

🜄 Hyperchromatic effect:

Chromophore increases absorption intensity (increases □)

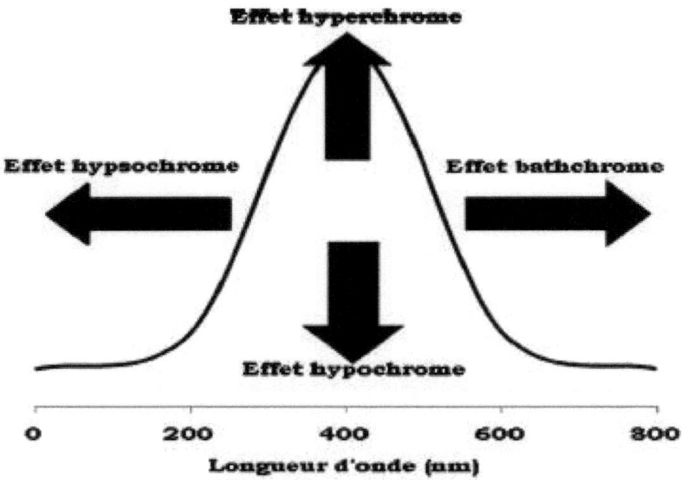

Figure 28: Effect of auxochrome grouping on absorption wavelength and intensity

IV. Examples of the effect of the environment on transitions

1. Effect of substitution

The position of the absorption band depends on the presence or absence of a substituent on the chromophore group. For example, the more the ethylenic group is substituted, the more the absorption band due to the $(\pi \rightarrow \pi^*)$ transition is shifted towards the visible: bathochromic effect.

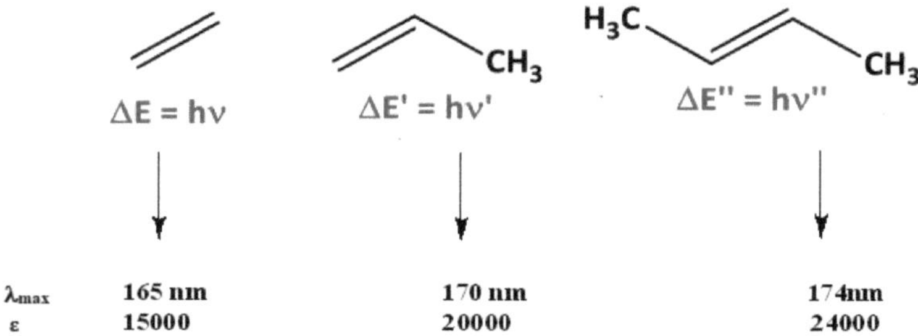

Figure 29: Bathochromic effect of a substituent on the ethylene group.

2. Effect of conjugation

• *Isolated chromophores*: for a series of molecules possessing the same chromophore, the position and intensity of the absorption bands remain essentially constant. If a molecule has several isolated chromophores, i.e. chromophores that do not interact with each other because they are separated by at least two single bonds, the effects of each individual chromophore are superimposed.

• *Chromophores in conjugated systems*: when chromophores interact with each other, the absorption spectrum is shifted towards longer wavelengths (*bathochrome effect*) with an increase in absorption intensity (*hyperchrome effect*). A special case is that of conjugated systems, i.e. organic structures with several unsaturated chromophores separated by a single bond. In this case, the spectrum is greatly disrupted compared to the simple superposition of the effects produced by the individual chromophores. The greater the number of carbon atoms over which the conjugated system extends, the smaller the gap between the levels of the frontier orbitals. This results in a very strong bathochromic effect

2.1. Ethylenic derivatives:

The sequence of unsaturations leads to the delocalization of π electrons. This delocalization, which reflects the ease with which electrons can move along the molecule, is accompanied by a convergence of energy levels.

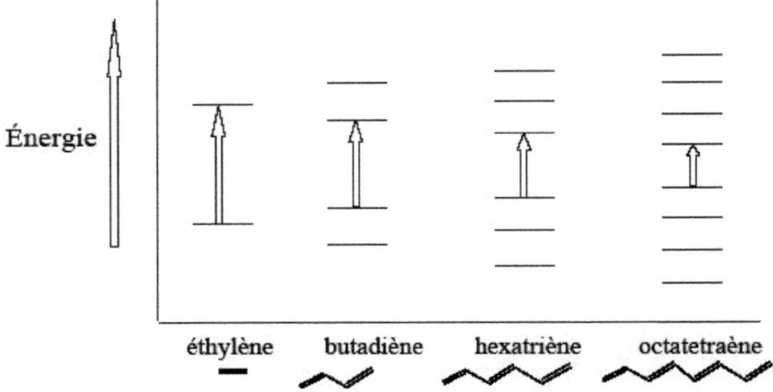

Figure 30: Electron delocalization effect π

Consider the following derivatives of the molecule

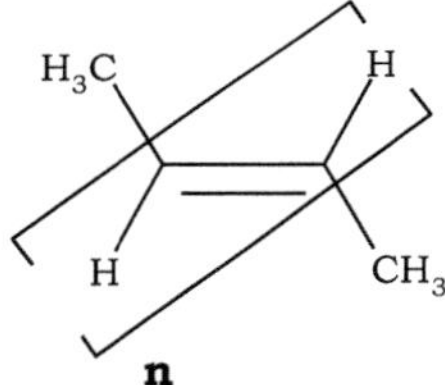

Figure 31: Polyene diagram

Table 6: The study focuses on the variation of wavelength and molar extinction coefficient as a function of the number n.

Number n	Wavelength λ (nm)	Extinction coefficient $\varepsilon.$ 10^4 (L/mol cm)
1	174	1,6
2	227	2,4
4	310	7,7
6	380	14,7

We can see that the wavelength and extinction coefficient increase as n increases. This is a double effect (bathchromic and hyperchromic).

The same effect is observed on the n$\rightarrow \pi^*$ transition.

Table 7: Increase in wavelength and extinction coefficient as n increases (double effect: bathchromic and hyperchromic)

Compounds	Transition $\pi \rightarrow \pi$	Transition $n \rightarrow \pi^*$
Propanone CH_3-CO-CH_3	188 nm	279 nm
4-Méthylpent-3-én-2-one or mesityl oxide	236 nm	315 nm

Note: Bathochromic displacement is responsible for the color of many natural compounds with extended conjugated chromophores.

β-carotene: The orange color of *β-carotene* comes from the union of eleven conjugated double bonds: $\underline{\lambda_{max} = \textbf{497 and 466 nm (in chloroform)}}$.

λ values$_{max}$ of a family of conjugated *E-disubstituted* polyenes that differ from each other in the number of conjugated double bonds. This effect is responsible for the color of many natural compounds whose semi-developed formulas feature extended conjugated chromophores. For example, the orange color of "all-trans" β-carotene (figure qq), results from the presence of eleven conjugated double bonds $\underline{(\lambda_{max} = \textbf{425, 448, and 475 nm in hexane})}$. The greater the electron delocalization, the greater the bathochrome effect.

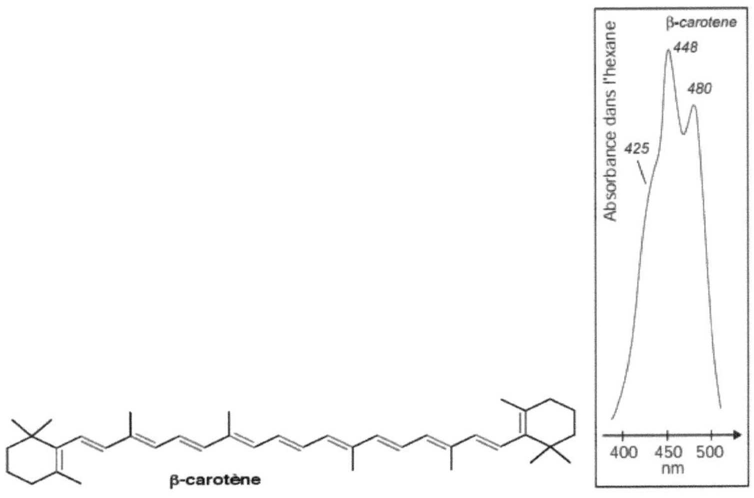

Figure 32: Effect of several conjugated double bonds on the position of the absorption maximum of the $\pi \to \pi^*$ transition for some conjugated polyenes.

The following molecules are assumed to have maximum wavelengths:

Z and E but-2-ene and **2-methyl-propene**: λ_{max} = **186** nm

2-methyl but-2-ene: λ_{max} = **191** nm

2,3-dimethyl but-2-ene: λ_{max} = **196** nm

The more the group is substituted, the more the absorption band is shifted towards □: bathchrome effect. An increment **of 5 nm per substituent is added**.

2.2. Benzene derivatives

Aromatic compounds lead to more complex spectra than ethylenics. The $\pi \to \pi^*$ transitions give rise to a "fine structure".

Benzene has 3 characteristic bands K, R and B, whose characteristics are given in the following table:

Table 8: Characteristic K, R and B bands for benzene

Band	Wavelength λ (nm)	Extinction coefficient ε (L/mol cm)
K	184	60 000
R	204	7400
B	254	204

By comparing the molar extinction coefficients of the 3 bands, we can say that band B is invisible compared to the other 2.

The effect of substitution of benzene by auxochrome groups on the variation of λ and ε especially of the invisible band B will be studied.

Table 9: The effect of benzene substitution with auxochrome groups on λ and ε

Compound	Auxochrome group	Wavelength λ (nm)	Extinction coefficient ε (L/mol cm)
Phenol	OH	270	1450

Aniline	NH_2	280	1430
Para-nitrophenol	NO_2	375	16000
Benzene		254	204

We note that wavelengths vary markedly from one auxochrome grouping to another. These groups produce a bathchromic effect. Absorption intensity also increases (hyperchromic effect) with these groups, and the best intensity was obtained with the NO group [2].

For polynuclear aromatics, as the number of condensed rings increases, absorption shifts to longer wavelengths until it reaches the visible region.

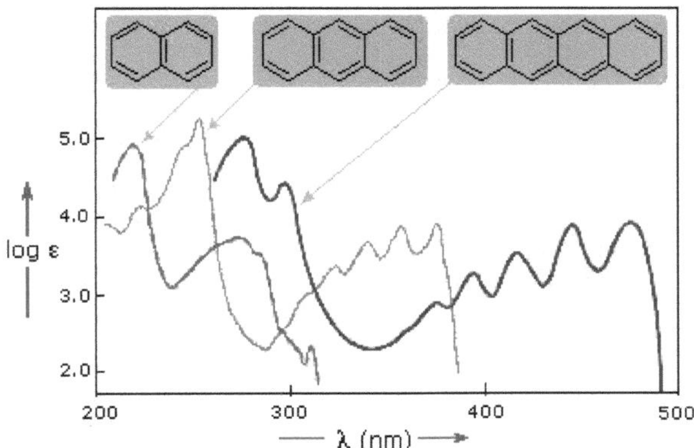

Figure 33: Effect of polynuclear aromatics

3. Solvent effects: Solvatochromy

3.1 Effect of pH

Assume addition of a base (NaOH) to an alcohol according to the reaction:

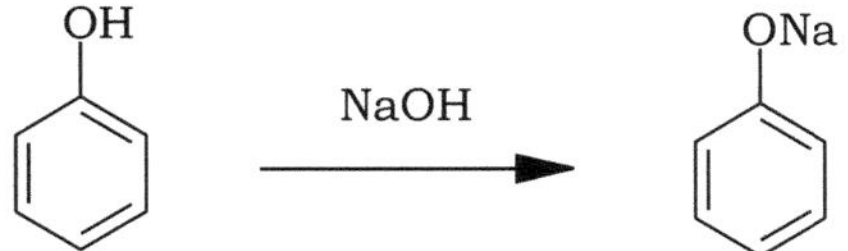

Figure 34: Reaction of NaOH with phenol

An increase in pH leads to a bathchromic and hyperchromic effect.
Carbon-carbon double bond

Table 10: The effect of increasing pH on λ and ε

—OH			—ONa		
Band	**λ (nm)**	**ε (L/mol cm)**	**Band**	**λ (nm)**	**ε (L/mol cm)**
R	211	6200	R	235	9400
B	270	1450	B	287	2600

The pH of the medium in which the analyte is dissolved can have a significant effect on the spectrum. Among the compounds that display this effect most dramatically are the colored indicators, whose color changes are used to advantage during acid/alkaline determinations (Figure 35). This is how equivalence points can be identified.

This compound is colorless at pH below 8 and bright pink at pH above 9.5. The graph (figure 35) shown here in perspective clearly shows that at acidic pH levels there is no absorption in the visible part of the spectrum. On the other hand, it's the absorption around 500 nm that appears when the pH becomes alkaline, which is responsible for the well-known color of this compound. In this example, note the pH-dependent modification of chemical bonds.

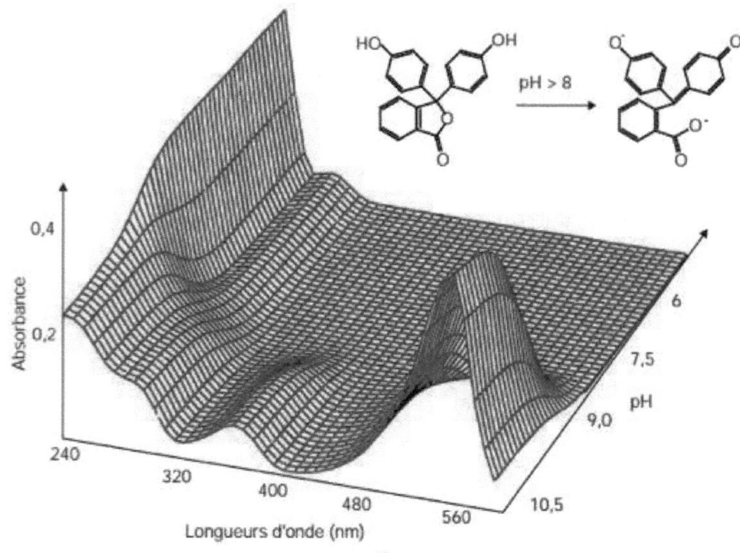

Figure 35: Effect of pH on a phenolphthalein solution.

3.2 Solvent effect: hypsochrome and bathochrome effects

For studies in solution, the solvent must be suitably chosen: it must dissolve the product and be transparent (non-absorbent) in the region under examination (table 11).

Table 11: The absorption zone for certain solvents and materials.

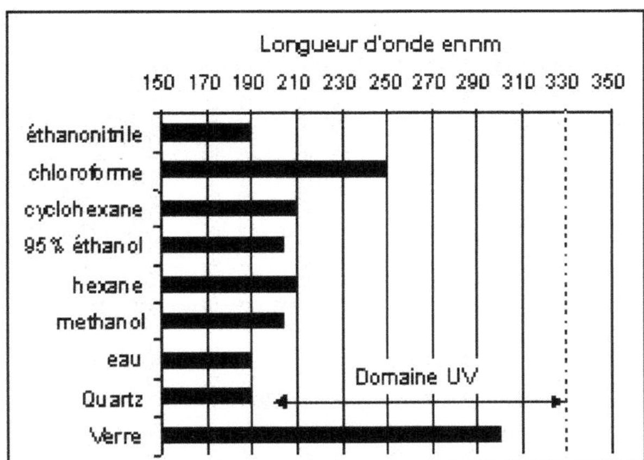

Example:

Hexane can be used as a solvent for samples that absorb at wavelengths above 210 nm.

Each solvent has its own polarity. Since we know that any electronic transition modifies the charge distribution in the compound in solution, it is obvious that the position and intensity of absorption bands will vary somewhat with the nature of the solvent used. These changes reflect the physical solvent-solute interactions that modify the energy difference between ground and excited states, and which are sufficiently clear to recognize the type of electronic transition in question. When solvent polarity is increased, two opposing effects can be distinguished.

3.2.1. Hypsochromic effect: Transition n $\pi^* \rightarrow$

If the chromophore responsible for the observed transition is more polar in its ground state than in its excited state, a polar solvent will predominantly stabilize the shape before absorption of the photon by solvation. More energy is therefore required to bring about the electronic transition in question, resulting in a shift of the absorption maximum towards shorter wavelengths than would occur in a non-polar solvent. This is the *hypsochrome effect*.

The same applies to the n$\rightarrow \pi^*$ carbonyl transition of ketones in solution. The C^+-O^- form (characterized by its dipole moment m) will be all the more stabilized the more polar the solvent. As the excited state is reached rapidly, the solvent cage surrounding the carbonyl does not have time to reorient to stabilize the situation after photon absorption. This same effect is observed for the $n \rightarrow \sigma*$ transition. It is accompanied by a variation in the coefficient

'

3.2.2. Bathochrome effect: Transition $\pi\rightarrow \pi$ *

For low-polarity compounds, the solvent effect is weak. However, if the dipole moment of the chromophore increases during the transition, the final state will be more solvated. A polar solvent will thus stabilize the excited form, which favors the transition: we observe a shift towards longer wavelengths, compared with the spectrum obtained in a non-polar solvent. This is the *bathochrome effect*. This is the case for the $\pi\rightarrow \pi$ * transition of ethylenic hydrocarbons whose initial double bond is not very polar.

Example: The bathochromic and hypsochromic effects of two different solvents are observed here, on both types of transitions.

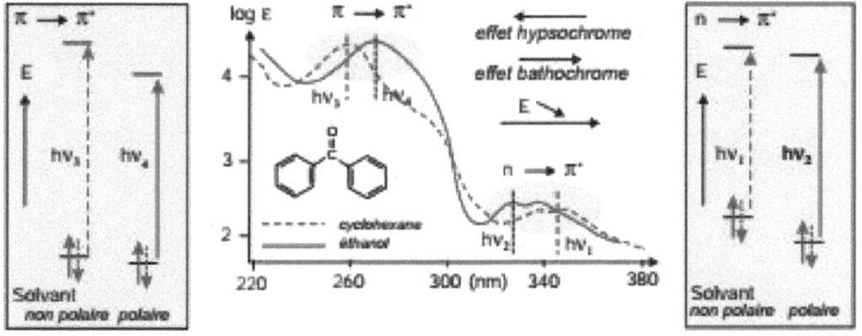

Figure 36: Spectra of benzophenone in cyclohexane **(1)** and ethanol **(2)**.

V. Application of Woodward-Fieser and Scott rules

1. Woodward-Fieser rules

1.1. Conjugated dienes

Basic structure	Increment added (nm)
Homo-annular diene	253
Hetero-annular or linear chain diene	214

Substituent	Increment added (nm)
Additional conjugated double bond	30
Exo-cyclic double bond	5
Alkyl or alkyl residue	5
-O-R	6
-S-R	30
-Cl, -Br	5
-NR$_2$	60
-O-CO-R	0

Example 1:

Basic structure	Homo-annular diene	Increment added (nm)
		253
Substituent	1 exocyclic double bond in relation to ring B	5
	3 alkyl residues in a, d and c	3 x 5 = 15
	A methyl in b	5
		278

max $\square$= 278 nm

1.2. carbonyl compounds α,β unsaturated

	X	max $\square$(nm)
	H	207
	Alkyle	215
	OH, Alcoxy	193

Special features of the structure:

If the molecule also has :

An exo-cyclic double bond is added in 5 nm increments.

For further conjugation, add 30 nm

A homo-annular diene component (open chain or cyclic) is added at 39 nm, taking substituent positions into account.

Incremental	α	β	γ	δ
Alkyle	10	12	18	18
Cl	12	12	***	***
Br	25	30	***	***
OH	35	30	***	50
Alcoxy	35	30	17	31
O-CO-R	6	6	6	6
NR$_2$	***	95	***	***

Note: These values are determined using Methanol or Ethanol as solvents.

<u>**Example:**</u>

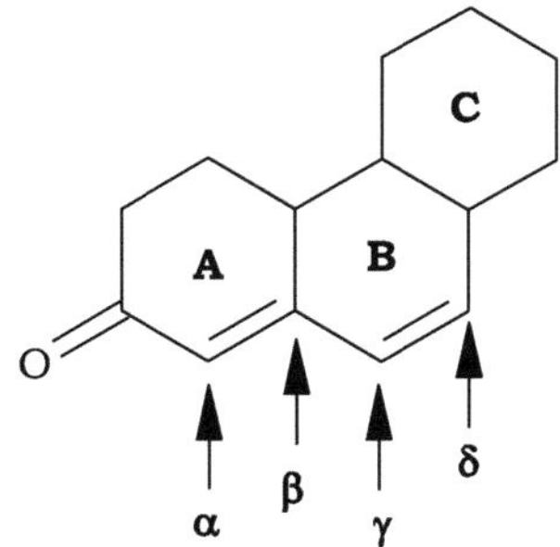

		Increment added (nm)
Basic structure	**Carbonyl compound α,β unsaturated**	215
Substituent	1 additional conjugated double bond	30
	1 exo-cyclic double bond	5
	1 remaining alkyl in β	10
	1 remaining alkyl in γ	18
		280

$_{max} \square = 280$ nm

2. Scott's rules

These are rules for calculating the maximum wavelengths of aromatic compounds.

Basic structure	**R**	**Base value (nm)**
	H	250
	Alkyl or Acyl	246
	O-alkyl	230
	OH	230

Benzene substitution increment in nm :

X	**ortho**	**meta**	**para**
Alkyl or cycle remainder	3	3	10
-OH, -OR	7	7	25
-Cl	0	0	10

-Br	2	2	15
-NHCOCH$_3$	20	20	45
-NR$_2$	20	20	85
-NH$_2$	15	15	58

Example 1:

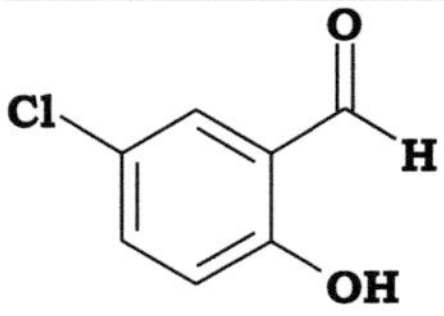	Basic structure	250 nm
	Chlorine in meta	0
	OH in ortho	7 nm
		257 nm

Example 2:

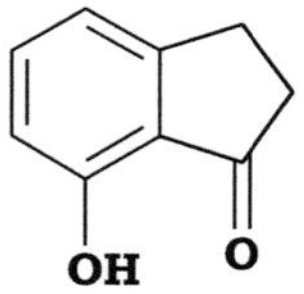	Basic structure	246 nm
	1 ortho alkyl residue	3 nm
	OH in ortho	7 nm
		256 nm

Example 3:

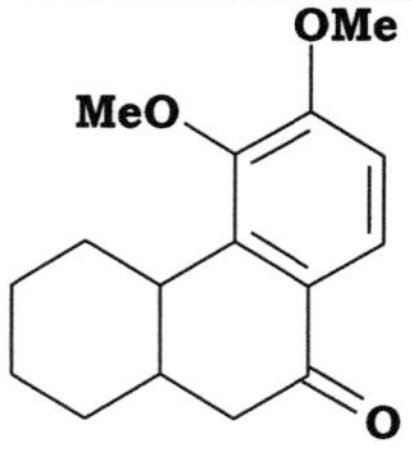	Basic structure	246 nm
	1 ortho alkyl residue	3 nm
	OCH$_3$ in meta	7 nm
	OCH$_3$ en para	25 nm
		281 nm

Note: Solvent corrections

Solvent	Correction (nm)
Water	+ 8
Chloroform	- 1
Ether	- 7
Cyclohexane	- 11

| Dioxane | - 5 |
| Hexane | - 11 |

VI. Equipment

1. Components of a UV-visible spectrophotometer

1.1 Composition of a molecular absorption spectrometer

The schematic diagram of a double-beam spectrophotometer is shown below:

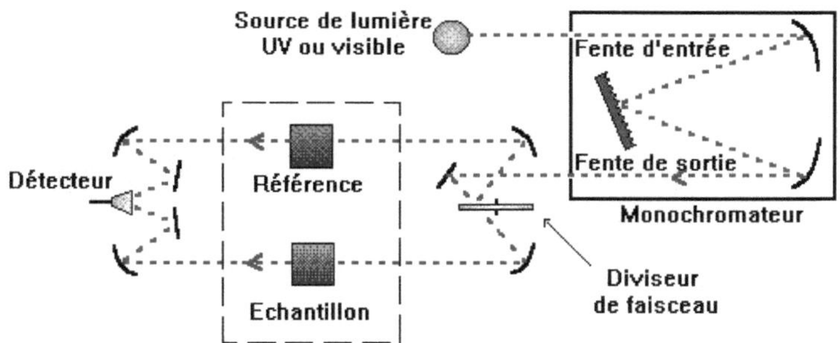

Figure 37: Principle of a double-beam spectrophotometer

A spectrophotometer consists of the following components:

<u>Light source</u>

It is made up of :

- A deuterium discharge lamp used in the 190 to 400 nm range, with maximum emission at 652.1 nm.

- A tungsten filament lamp for the 350 to 800 nm region

- A xenon discharge lamp used in the UV and visible range. This type of lamp is highly energetic. It operates as a flash, just when you need to take a measurement.

<u>Monochromator</u>

The basic element is a prism, grating or colored filter. The role of the monochromator is to isolate the radiation being measured. Its main components are a dispersive system, an input slit and an output slit.

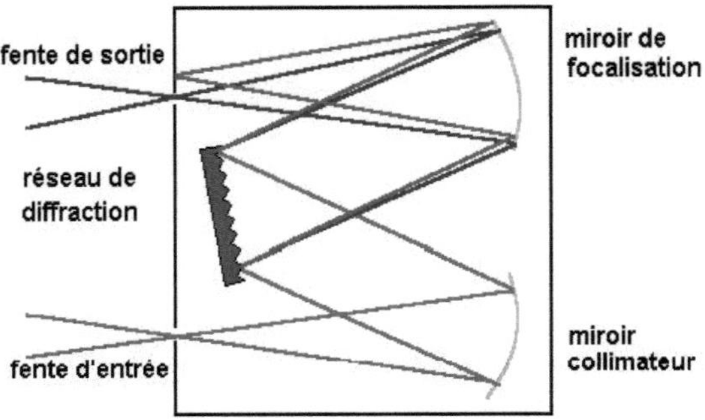

Figure 38: Monochromator

Tank

It contains either the sample or the reference. The length of the cell is defined (1, 2, 4 or 5 cm optical path). It must be transparent to the study radiation. For example, UV cuvettes are made of quartz, and can be neither glass nor plastic.

Detector

Photodiode (semiconductor)

When a photon encounters a semiconductor, it can transfer an electron from the valence band (low energy level) to the conduction band (high energy level), creating an electron-hole pair. The number of electron-hole pairs is a function of the amount of light received by the semiconductor, which can therefore be used as an optical detector.

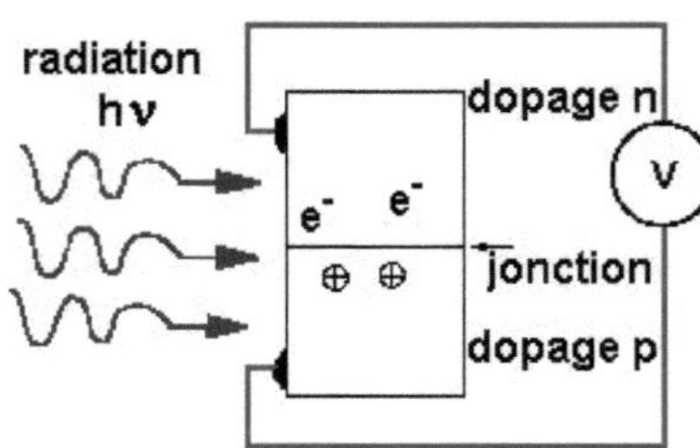

Figure 39: Detector

Diode array

The use of a diode array allows simultaneous measurement across the entire spectrum. A CCD array is an alignment of small photodiodes (14µm x 14 µm) that function as a light integrator. The charge appearing in a photodiode is proportional to exposure, i.e. the product of illumination and exposure time, and is wavelength-dependent. At the end of the exposure, the sensor content is transferred to an analog shift register, and a new exposure begins. This register transmits the stored data in serial mode, i.e. one after the other at a rate set by the CCD array's control electronics. This data appears as a voltage. In the Mecacel spectrophotometer, these voltages are converted into a table of numbers by the interface linking the spectrophotometer to the computer. The software processes this table of values. Coupled to a computer, the spectrophotometer can be used to plot absorption spectra very quickly. The software manages the exposure time of the CCD sensor.

Photomultiplier

Incident radiation photo-electrically pulls an electron from the cathode. This electron is then accelerated towards a second electrode, the dynode, raised to a higher potential. The energy of the incident electron is sufficient to pull out several other electrons and so on, hence the multiplicative effect. For every electron stripped from the cathode, up to 106 electrons can be recovered from the anode.

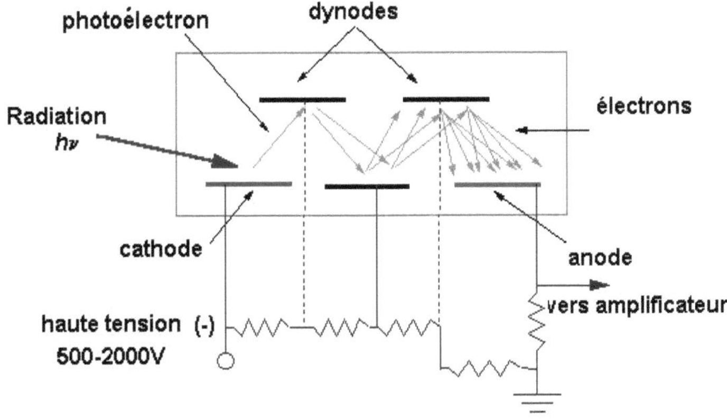

Figure 40: Photomultiplier

VII. Applications of UV-Visible spectroscopy

Qualitative analysis

UV spectra generally provide little information on the molecular structure of compounds. Nevertheless, they are used either for confirmation or for identification using rules of thumb.

Quantitative analysis

Quantitative analysis by UV-Visible spectrometry is much more widely used than qualitative analysis, thanks to the use of the Béer-Lambert law.

Applications include :

Determination of iron in water or medication

Determination of active molecules in a pharmaceutical preparation

Determination of benzene in cyclohexane

Other applications

Other applications include quality control or monitoring reaction kinetics, determining acid dissociation constants or complexation constants...

Appendices: UV spectroscopic data tables

APPENDIX 1: Visible wavelength and color table

Longueur d'onde de la radiation absorbée (nm)	Couleur perçue	Couleur de la radiation absorbée
400-435	jaune-vert	violet
435-480	jaune	bleu
480-490	orangé	vert-bleu
490-500	rouge	bleu-vert
500-560	pourpre	vert
560-580	violet	jaune-vert
580-595	bleu	jaune
595-625	vert-bleu	orangé
625-800	bleu-vert	rouge

APPENDIX 2: Table of some visible wavelength and color values

Chromophore (dans UV)	Transition	λ_{max} / nm)	$\log(\varepsilon_{max})$	Polyène conjugué	Formule	λ max	$\log(\varepsilon_{max})$	Couleur
Alcène C=C	π-π^*	175	3,0	Ethène	C_2H_4	162	4,0	Incolore
Alcyne C≡C	π-π^*	180	1,5	Benzène	C_6H_6	180	4,8	Incolore
Aldéhyde - Cétone C=O	π-π^* n-π^*	180 280	3,0 1,5	Buta-1,3-diène	C_4H_6	217	4,3	Incolore
Acide carboxylique COOH	n-π^*	205	1,5	Hexa-1,3,5-triène	C_6H_8	258	4,5	Incolore
Nitro NO₂	n-π^*	271	< 1,0	Octa-1,3,5,7-tétraène	C_8H_{10}	296	4,7	Incolore
				-	$C_{10}H_{12}$	335	5,1	Jaune pâle
				-	$C_{16}H_{18}$	415	5,3	Orange
				-	$C_{22}H_{24}$	470	5,3	Rouge

APPENDIX 3: <u>UV</u> SPECTROSCOPIC DATA TABLE
Woodward-Fieser rules: λ_{max} prediction for conjugated dienes in ethanol

homo-annulaire	hétéro-annulaire	en chaîne linéaire
253 nm	214 nm	217

Substituant	Incrément ajouté (nm)
Conjuguée supplémentaire	30
Double liaison exo-cyclique	5
Alkyle ou reste alkyle	5
-Cl, -Br	5
-OR	6
-S-R	30
-NR$_2$	60
-O-CO-R	0

APPENDIX 4: <u>UV SPECTROSCOPIC DATA TABLE</u>
<u>Woodward-Fieser rules: λ_{max} prediction for carbonyl compounds α,β-unsaturated in ethanol</u>

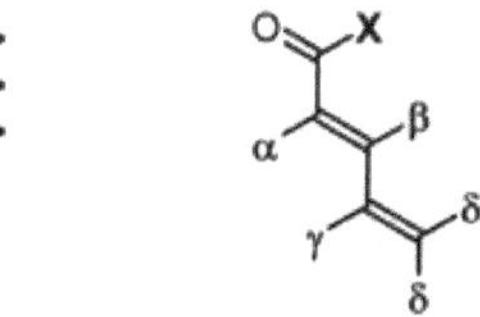

Incréments	α	β	γ	δ
Alkyle	10	12	18	18
Cl	12	12	***	***
Br	25	30	***	***
OH	35	30	***	50
Alcoxy	35	30	17	31
O-CO-R	6	6	6	6
NR$_2$	***	95	***	***

- **X= H** λ**max = 207nm**
- **X= Alkyle** λ**max = 215 nm**
- **X= OH, Alcoxy** λ**max=193 nm**
- C=C exo-cyclique, **5 nm**
- Une conjugaison supplémentaire,**30 nm**
- Diène homo-annulaire **39 nm** en tenant en compte des positions des substituants.

APPENDIX 5: <u>UV</u> SPECTROSCOPIC DATA TABLE

<u>Scott's rules: λ_{max} prediction of aromatic carbonyl compounds in ethanol</u>

Structure de base	R	Valeur de base (nm)
	H	250
	Alkyle ou Acyle	246
	O-alkyle	230
	OH	230

X	ortho	méta	para
Alkyle ou reste de cycle	3	3	10
-OH, -OR	7	7	25
-Cl	0	0	10
-Br	2	2	15
-NHCOCH$_3$	20	20	45
-NR$_2$	20	20	85
-NH$_2$	15	15	58

APPENDIX 6: <u>UV</u> SPECTROSCOPIC DATA TABLE
<u>Solvent corrections</u>

Solvent	Correction (nm)
Water	+ 8
Chloroform	- 1
Ether	- 7
Cyclohexane	- 11
<u>Dioxane</u>	- 5
<u>Hexane</u>	- 11

Chapter III: Infrared (IR) spectroscopy

I. Infrared (IR) spectrophotometry concepts

This technique focuses on the vibrations of bonds between atoms within a molecule. It is absorption spectroscopy.

1- Infrared radiation

Infrared (IR) radiation was discovered in 1800 by Frederick Wilhelm Hershel. This radiation, located beyond red wavelengths, lies between the visible spectrum and Hertzian waves.

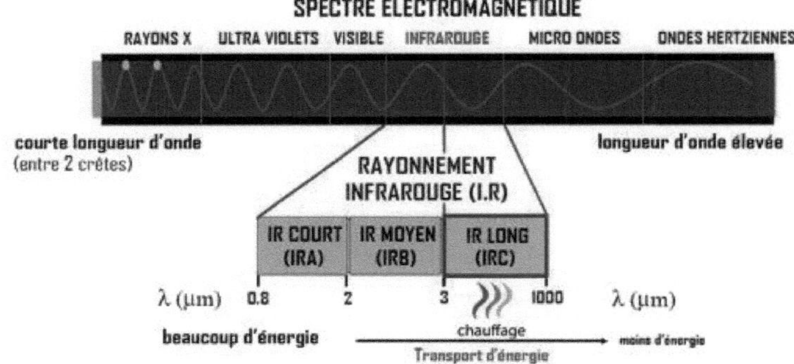

Figure 41: Schematic representation of the electromagnetic spectrum

The infrared range extends from 0.8 µm to 1000 µm. It is arbitrarily divided into 3 categories (Figure 42):

- Near infrared (0.8 to 2.5 µm or 12500-4000 cm), [-1]

- The mid-infrared (2.5 to 25 µm or 4000-400 cm[-1]) and the infrared (2.5 to 25 µm or 4000-400 cm).

- The far infrared (25 to 1000 µm or 400-10 cm).[-1]

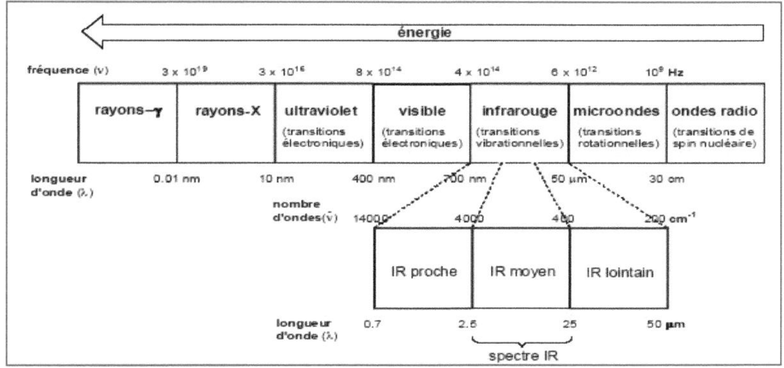

Figure 42: *IR domains in the electromagnetic spectrum*

Later (in 1924), it was discovered that the energy of average infrared radiation coincides with that of the molecule's internal motions. This suggests a relationship between a molecule's absorption of infrared radiation and its molecular structure.

Infrared radiation easily penetrates the atmosphere. This property is used in aerial photography, to take panoramic views on overcast days. Infrared is also used for domestic and industrial heating, drying varnish and paint, wood, leather, paper and photographic film, and dehydrating fruit and vegetables. One of the most important military applications is infrared homing of missiles. In therapy, infrared rays activate cellular processes, particularly healing.

IR spectroscopy is widely used to determine the functional groups of a molecule.

The movements of atoms in a molecule can be classified into three categories:

- translations

- rotations

- vibrations

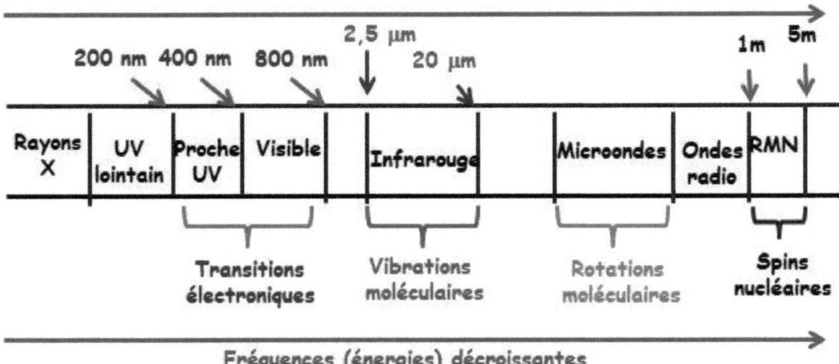

Figure 43: The three categories of atomic motion in a molecule

The mechanical energy of an isolated molecule results from the combination of three independent quantized terms corresponding to its rotational energy E_R, vibrational energy E_V and molecular electronic energy E_E. The values of these energies are very different from one another.

$E_T = E_R + E_V + E_E$

An electromagnetic wave of frequency ν_0 can be absorbed by a molecule, which will move from one energy level to another.

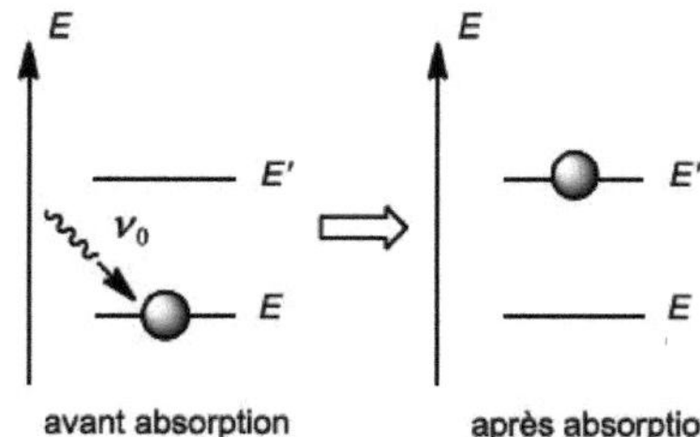

Figure 44: Schematic representation of the absorption of an electromagnetic wave

Absorption is only possible if the energy of the wave corresponds to the energy difference between the two energy levels:

$$E' - E = h\nu_0 = \frac{hc}{\lambda_0}$$

Figure 45: Energy difference between the two energy levels

Electronic transitions take place in the UV-visible range.

Vibrational transitions take place in the infrared range

UV-visible radiation → electronic transition

IR radiation → bond vibration

Since most samples are condensed liquids or solids, rather than isolated species, there are many dipole-dipole interactions which perturb the energy levels, and therefore the absorption wavelengths. As a result, we are always faced with spectra made up of broad peaks called bands, which can extend over tens of cm^{-1} and which no instrument can break down into individual transitions.

Vibrational spectroscopy studies molecular vibrations with transitions in the range (30 - 13,000 cm^{-1}). The energy of photons in the infrared modifies both E_R and E_V . The molecule becomes a rotator-oscillator, which in turn gives a molecular rotation-vibration spectrum.

II. Study of the vibration of a diatomic molecule

1. Harmonic oscillator model

1.1- Principle

Infrared energy (produced, for example, by an incandescent filament) lies between 20 μm and 2.5 μm for the spectra commonly used by organics, or between 500 cm^{-1} and 4,000 cm^{-1} for wave numbers, corresponding to energies of around 30 kJ-mol $.^{-1}$

Such energy will not be sufficient to enable an electronic transition as in visible UV spectroscopy, but a scan between these wavelengths will act on the vibration and rotation of molecules.

Molecules are not rigid, and the atoms that make them up can vibrate in relation to each other, constituting vibrators. To model bond vibrations, we refer to the harmonic oscillator. Two atoms A and B united by a covalent bond are likened to two masses *mₐ and m_B* respectively *of an A-B molecule,* linked by a spring with stiffness constant k. *If we call k the* spring constant of this "spring", then the frequency ν is given by the classical Hooke's law demonstrated in elementary physics lessons:

Diatomic molecule≡ harmonic oscillator:

    * mass μ (reduced mass)

    * force constant k

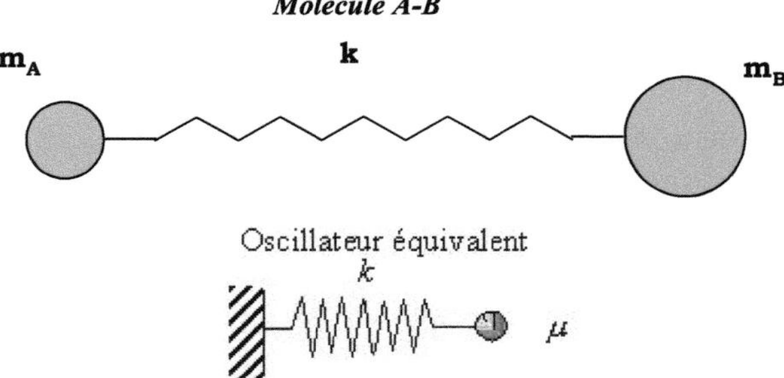

Figure 46: A-B diatomic molecule or equivalent harmonic oscillator

Frequency of oscillation by Hooke's law: Masses can oscillate around their equilibrium position with a frequency given by Hooke's law:

$$\upsilon = \frac{1}{2\pi}\sqrt{\frac{k}{\mu}} \, ,$$

With:

k: Stiffness constant (bond strength constant) (N m)$^{-1}$

μ: reduced mass, calculated by the following expression:

$$\frac{1}{\mu} = \frac{1}{m_A} + \frac{1}{m_B}$$

With m_A and m_B being the masses of atoms A and B

$$\mu = \frac{m_A m_B}{m_A + m_B}$$

Remembering that the wave number is given by the formula:

$$\bar{v} = \frac{1}{\lambda} = \frac{v}{c}$$

Figure 47: Frequency given by Hooke's law

Note 1:

When this diatomic molecule is subjected to the action of an electromagnetic wave characterized by the frequency v_0 , this radiation is absorbed (resonance phenomenon) when $v = v._0$

Note 2:

The wave number σ is the inverse of the wavelength λ expressed in cm $_{-1}$

Note 3:

> The practical quantity in vibrational spectroscopy is the wave number.

$$\bar{v}\,[cm^{-1}] = \frac{1}{2\pi c}\sqrt{\frac{k}{\mu}}$$

Figure 48: Wave number given by Hooke's law

The Hooke frequency or vibrational frequency of a covalent bond depends on the reduced mass (mass of the atoms) and the stiffness constant (nature of the bond).

In infrared spectroscopy, an absorption wavelength will most often characterize a function.

The following table gives an order of magnitude of the Hooke frequencies and the associated wave numbers for some covalent bonds.

Table 12: Reduced mass μ, wave number □ □et stiffness constant **k** of some covalent bonds

Link	k (N m)$^{-1}$	μ (Kg)	$\overline{v}$ (cm)$^{-1}$
C-C	145 - 900	$9{,}96.10^{-27}$	640 - 1600
C=C	970	$9{,}96.10^{-27}$	1650
C-O	400 - 700	$1{,}14.10^{-27}$	1000 - 1300
C=O	1200	$1{,}14.10^{-27}$	1720

2. Rotation spectrum - vibration

We assume that the two motions of rotation and vibration are independent of each other (rigid rotator and harmonic oscillator). Under these conditions, each motion retains its own characteristics:

<u>Energy levels</u>

<u>Quantum numbers</u>

<u>Selection rules</u>

2.1. Vibrational energy

This model can be used at the scale of the chemical bond, provided that the quantum aspect governing species at atomic dimensions is brought into play. A bond whose vibrational frequency is v, will be able to absorb light radiation provided its frequency is identical.

According to this theory, the possible values of vibrational energy (E_V) are given by the following expression:

$$E_V = h\upsilon\left(v+\frac{1}{2}\right)$$

With:

v: Vibration quantum number (natural integer)

h: Planck's constant ($6.625.10^{-34}$ j.s)

The vibrational energy E_V is also expressed as:

$$E_V = \frac{h}{2\pi}\sqrt{\frac{k}{\mu}}\left(v+\frac{1}{2}\right) = \hbar\omega_0\left(v+\frac{1}{2}\right)$$

With:

$$\omega_0 = \sqrt{\frac{k}{\mu}}$$ Pulsation associated with the vibration frequency

Figure 49: Vibrational energy

Remarks:

* At room temperature, molecules are in a non-excited state ($v = 0$).

* All vibrational energies can be calculated by assigning appropriate values to the vibrational quantum number.

2.2 Rotational energy

The rotational energy E_R is expressed as :

$$E_R = \frac{h^2}{8\pi^2 I} J(J+1)$$

With :   J: Rotation quantum number (natural number)

I: Moment of inertia

Figure 50: Rotational energy

Remarks:

* At room temperature, molecules are in a non-excited state ($J = 0$).

* The set of rotational energies can be calculated by giving appropriate values to the rotational quantum number.

2.3 Selection rules

Vibrational and rotational transitions are permitted if they comply with the selection rules ($\Delta v = 0, \pm 1$) and ($\Delta J = \pm 1$) :

$\Delta v = 0$: pure rotation

$\Delta v = -1$ and $\Delta J = \pm 1$: the spectrum obtained is an emission spectrum.

$\Delta v = +1$ and $\Delta J = \pm 1$: the spectrum obtained is an absorption spectrum.

2.4. Fine structure

It is assumed that the molecule in the ground state is characterized by the quantum constants J and v. In the excited state, it is characterized by J' and v'. Whereas in the excited state, it is characterized by J' and v'.

When absorbing radiation, selection rules dictate that:

$\Delta v = +1$ and $\Delta J = \pm 1$: the spectrum obtained is an absorption spectrum.

The spectrum structure comprises two series of lines or two branches with the following notations:

D → D+1: Branch R (Rich)

J → J-1 : Branch P (Rich)

The forbidden transition $J = 0 \rightarrow J' = 0$ (pure vibration) corresponds on the spectrum to the Q branch. Possible transitions are shown in the following energy diagram:

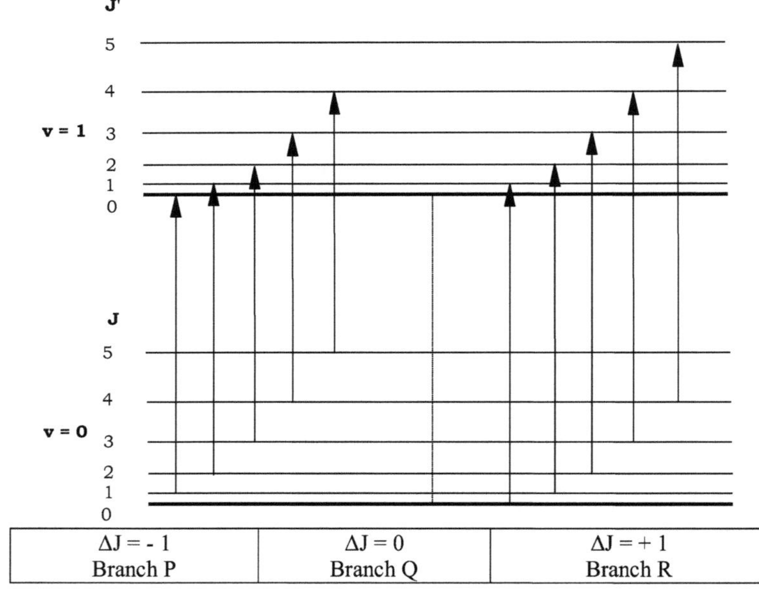

Figure 51: Energy diagram of possible transitions

2.5. Rotation - vibration spectrum

The above rotation-vibration transitions lead to a series of lines represented as follows:

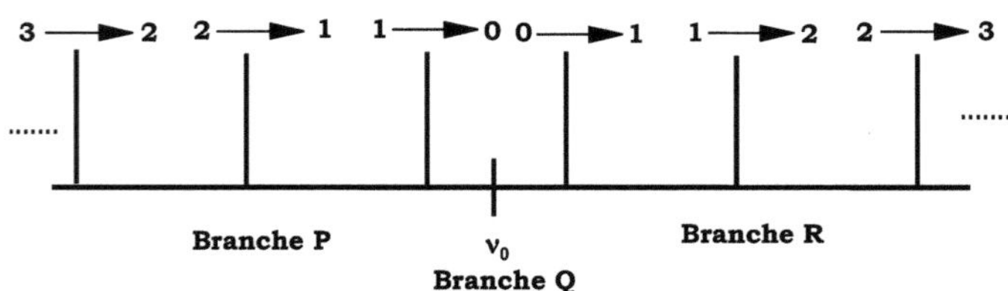

Figure 52: Rotation-vibration spectrum

Please note:

The heights of the lines are not equal; their intensities depend on the populations of the rotational levels concerned, which are a function of the values of the rotational quantum number J. In fact, a vibration-rotation spectrum has the following appearance:

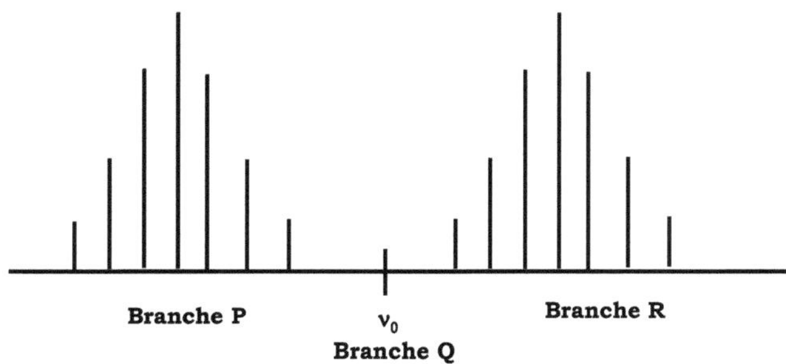

Figure 53: Example of a rotation-vibration spectrum of gaseous HCl

3. Parameters influencing wave numbers

3.1 Influence of bond strength

The first step is to compare the nature of the bond with the Hooke frequency. As much as the bond is stiffer (k is larger), the vibration frequency increases (wave number also increases).

<u>Effect of k:</u>

Vibration frequency proportional to k;

For example:

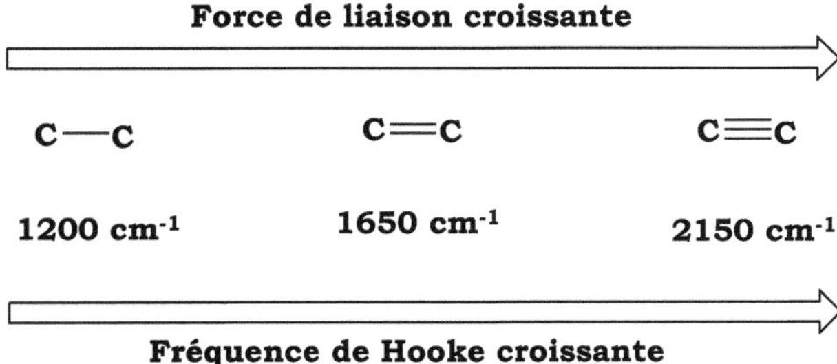

Figure 54: Effect of k on vibration frequency (frequency proportional to k)

<u>**Effect of carbon hybridization state:**</u>

On the other hand, the Hooke frequency of a C-H bond depends on the hybridization state of the carbon. Indeed, the Hooke frequency is high if the carbon is sp-hybridized .[3]

For example:

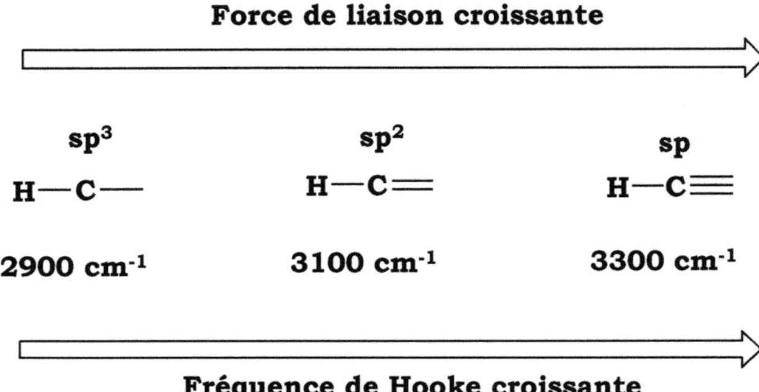

Figure 54: Effect of carbon hybridization state on vibration frequency

3.2 Influence of atomic masses

An example is given below where the mass of the atoms will have an effect on the Hooke frequency.

<u>**Effect of μ:**</u>

Vibration frequency inversely proportional to μ:

For example:

Table 13: Effect of μ on vibration frequency ($\overline{V}$ inversely proportional to μ)

Link	m_2 (g)	μ (g)	$\overline{V}$ (cm)⁻¹
C-H	1	0,923	3300
C-C	12	6,000	1200
C-O	16	6,857	1100
C-Cl	35,5	8,968	800
C-Br	80	10,43	550

C-I	127	10,96	500

3.3 Influence of link polarization

Only vibrations involving a variation in the molecule's dipole moment are observed in the infrared. As a result, the vibration of polarized bonds will give rise to intense bands, while the bands of non-polarized bonds will be barely visible, if at all.

Figure 54: Effect of polarization on band intensity

III. Vibrations of polyatomic molecules

In the case of polyatomic molecules, the number of bonds increases and bond geometry becomes complex. According to vibrational theory, a molecule containing N atoms has 3N-6 vibrational degrees of freedom, and 3N-5 for linear molecules.

Only bonds with an oscillating dipole electric moment are active in the infrared.

Each vibrational mode has its own fundamental frequency and several other frequencies associated with harmonics.

In addition, interactions between the vibrational modes of a particular bond and those of other bonds are often observed. These interactions result in the appearance of combination bands.

1. Normal vibration mode

A normal mode of vibration is one in which all the atoms in the molecule vibrate at the same frequency and pass through their equilibrium position simultaneously. During vibration, the molecule's center of gravity remains unchanged.

2. Normal vibration mode types

2.1. Elongation (or valence) vibration

This is the vibration between two identical or different atoms. It depends on the masses of the atoms and the strength of the bond. It can be symmetrical or asymmetrical.

Example: Vibrational mode of elongation of a group ($-CH_2-$).

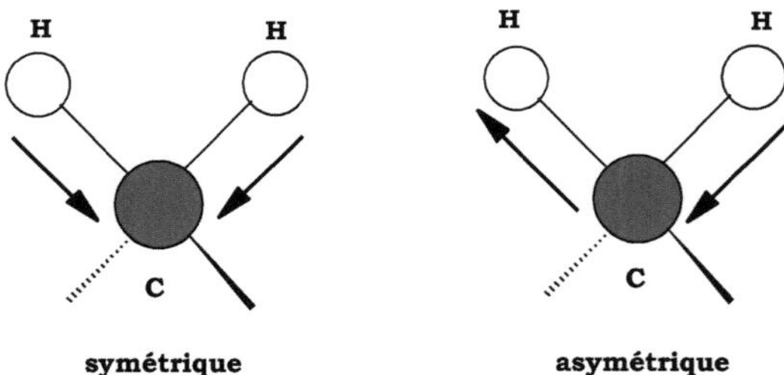

Figure 55: Elongation vibration mode of a group ($-CH_2-$)

2.2. Angular deformation vibration

This is the angular variation between two links.

Example: Vibrational mode of angular deformation of a ($-CH_2-$) cluster in the plane:

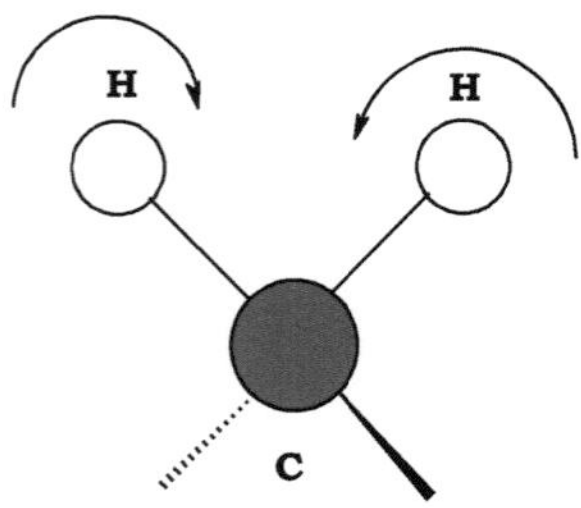

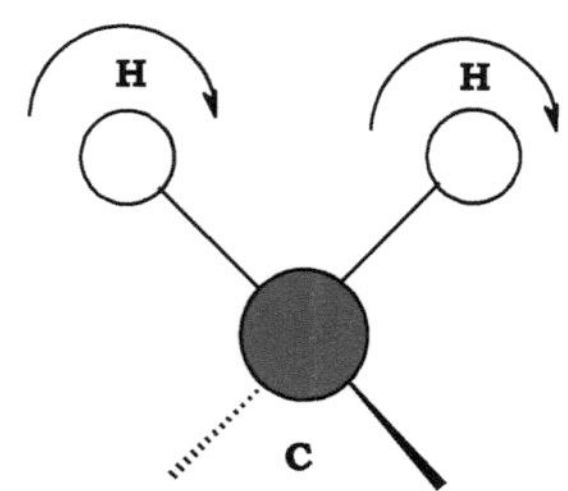

Figure 55: Vibrational mode of angular deformation of a (-CH$_2$ -) cluster in the plane

Out of plan:

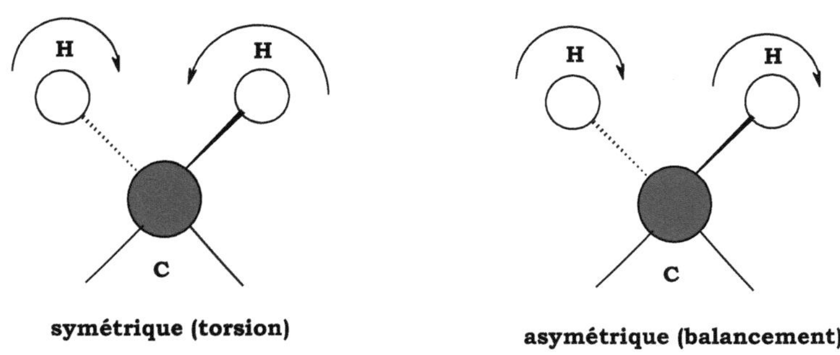

Figure 56: Angular deformation vibration mode of a (-CH$_2$ -) cluster

2.3 Characteristic frequencies

A frequency is said to be characteristic when it is present in all compounds containing the same structural element.

For example, the valence vibrations of the C— H bonds in compounds containing CH$_3$, CH$_2$, CH groups have characteristic frequencies specific to these groups, the frequencies and expected intensities of which are shown in the table below:

Table 14: Intensities and characteristic frequencies of the bands linked to the CH$_3$, CH$_2$, CH

Group	Vibration	Frequency	IR intensity
— CH$_3$	v_a(CH)$_3$	2962± 10 cm^{-1}	Strong
— CH$_2$	v_a(CH)$_2$	2926± 10 cm^{-1}	Strong
— CH	v(CH)	2890± 10 cm^{-1}	low
— CH$_3$	v_s(CH)$_3$	2872± 10 cm^{-1}	Strong
— CH$_2$	v_s(CH)$_2$	2853± 10 cm^{-1}	Strong

The IR spectrum of n-octane clearly shows the 4 absorptions corresponding to the valence vibrations of the CH_3 and CH_2 groups (figure 57).

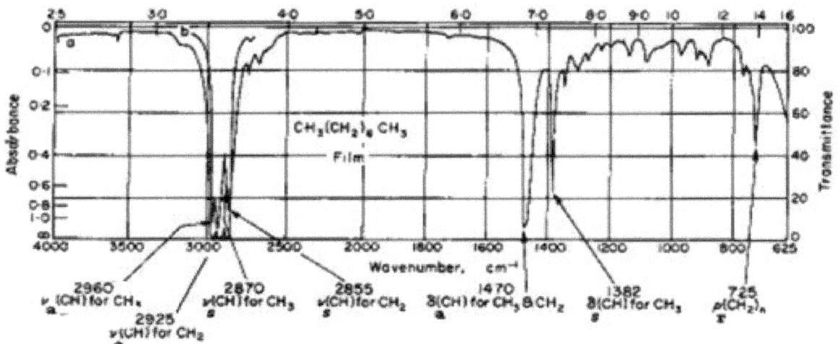

Figure 57: Infrared spectrum of *n-octane* in the liquid state

3. Active or inactive vibrations - degenerate vibrations

➤ Not all vibration movements are active in infrared.

➤ It's important to note, however, that a bond with zero dipole moment will give no signal in infrared spectroscopy. Thus, dioxygen O=O will have no active infrared bands, while ethylene will show signals due to C-H bonds, but the C=C double bond will not be active, as its dipole moment is zero due to the symmetry of the molecule.

➤ Active vibration in IR spectroscopy: dipole moment variation

➤ Doubly or triply degenerate modes: Modes with the same vibration frequency

Homo-nuclear molecules, such as H_2 , O_2 (see figure 58), have no dipole moment, and whatever the internuclear distance, they do not interact with the oscillating electric field.

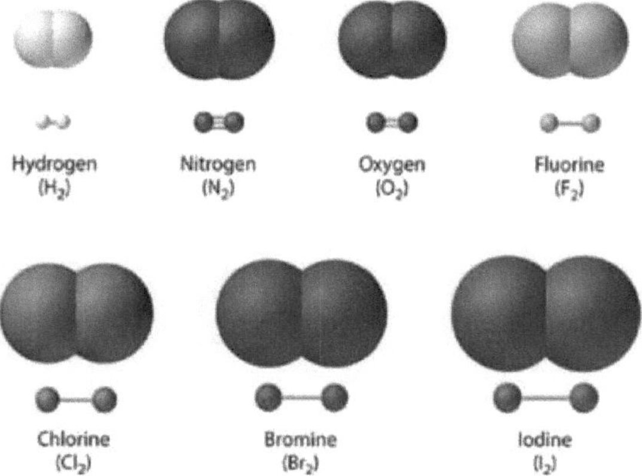

Figure 58: List of diatomic elements.

4. IR spectrum

➢ IR spectrum = sequence of absorption bands, more or less broad

➢ Abscissa: cm^{-1} and ordinate: Transmittance (%T) or Absorbance A = log(1/T)

Let's not forget that the IR spectrum of an organic molecule, with its many bonds, is complex and generally cannot be fully interpreted.

Typically, IR spectra are recorded with the inverse of the wavelength λ expressed in cm, or wavenumber σ, as the abscissa; λ expressed varies from 25 to 2.5 μm, σ varies between about 400 and 4000 cm .$^{-1}$

$$\lambda = \frac{c}{\nu} \qquad \frac{1}{\lambda} = \frac{\nu}{c} = \sigma$$

On the ordinate, the transmittance T, or its percentage, is plotted for each radiation:

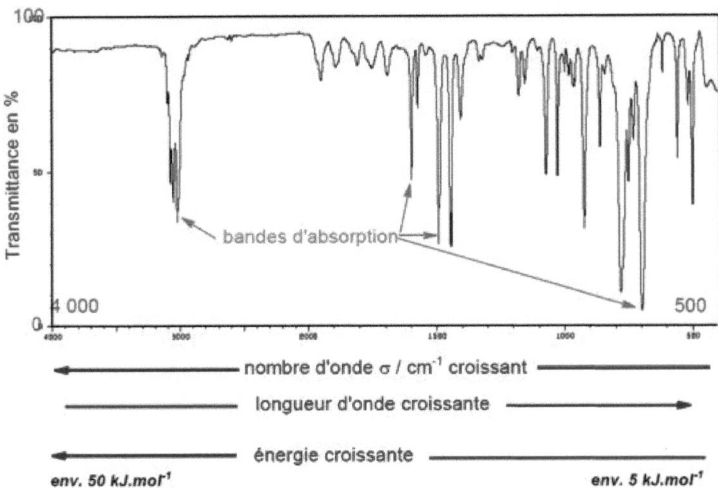

Figure 59: Infrared spectrum pattern

➤ Some fundamental vibrations are absent from the IR spectrum due to their <u>inactivity</u>.

➤ But <u>other</u> absorption <u>bands also</u> appear.

For each vibrational transition, a multitude of rotational transitions are induced by the excitation light, giving the peak of the vibrational transition the appearance of an absorption band (Figure 60).

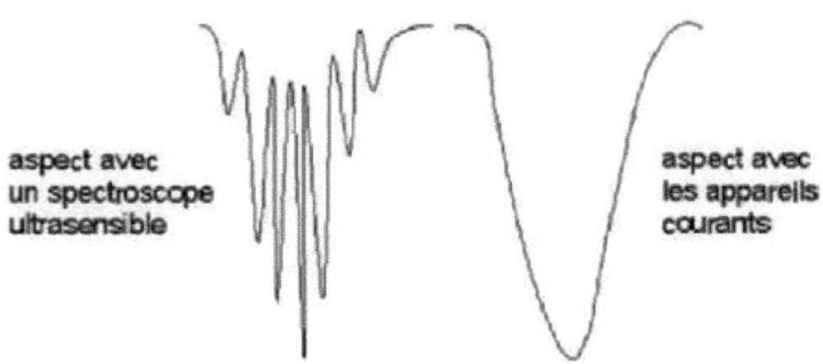

Figure 60: Appearance of absorption bands.

5. Types of vibration movements and notations.

For a diatomic molecule (CO, HCl...), the only possible vibration is the valence vibration, i.e. the one that takes place along the bond.

The problem becomes even more complicated in the case of polyatomic molecules, as deformation vibrations are added to the valence vibrations already present in diatomic molecules. Deformation vibrations require less energy than valence vibrations.

6. Consequences

An infrared spectrum is obtained by placing the sample in a vessel, usually made of NaCl or KBr. The IR spectrum generally indicates transmittance as a function of wave number between 4000 and 200 cm $.^{-1}$

Transmittance is equal to the percentage of incident radiation that has passed through the measuring cell. Wave number (expressed in cm^{-1}) is equal to the inverse of wavelength.

To interpret such a spectrum, we use tables indicating the characteristic absorption ranges of different chemical functions.

In the scanning range, it is customary to distinguish essentially three main regions in an IR spectrum:

First region (4,000 cm^{-1} - 1,500 cm^{-1}): Functional zone, where the bands corresponding to the vibrational transitions of most functional groups (R-C-H, R-OH, R-N-H, R-CH=O, RCO-R, R-COOH,.......) are found.

Second region (1500 cm^{-1} - 1000 cm^{-1}): Fingerprinted, this is a region of numerous small bands corresponding to the vibrational transitions of deformation. This region is totally characteristic of the molecule.

Third region (1000-400 cm-1): This is a low-energy region, where out-of-plane deformation transitions of the C-H bonds of alkenes and aromatics are predominant. This is in fact a less important region than the previous two.

The table below shows some IR bands representing significant valence vibrations.

Table 15: Some important valence vibrations

Links	Frequency range (cm)$^{-1}$	Intensity[a]
$-\overset{\shortmid}{\underset{\shortmid}{C}}-H$	2850-2980	m-f

	3010-3040	m
$=\overset{\displaystyle |}{C}-H$	~3330	f
$-O-H$	3550-3450 (free OH) 3100-3550 (OH bonded)[c]	m-f
$-N-H$	3300-3500	m
$-\overset{\displaystyle |}{\underset{\displaystyle |}{C}}-O-$	1000-1260	m
$-\overset{\displaystyle |}{\underset{\displaystyle |}{C}}-N\diagdown$	1020-1360	m
$\diagup\!\!\!C=C\!\!\diagup$	1650-1675	fa-m
$\diagdown C=O$ [b]	~1660 (amide) 1705-1725 (ketone) 1720-1740 (aldehyde) ~1750 (ester) ~1760 (carboxylic acid)	f f f f f
$-C\equiv C-$	2100-2260	fa-m
$-C\equiv N$	2200-2400	m

[a] fa = weak, m = medium, f = strong

[b] energies (i.e. wave numbers) can be lowered with conjugations

[c] engaged in a hydrogen bond

7. Group vibrations

Groups within a molecule (C=O, C-O, O-H, C-N, N-H, etc.) can be excited almost independently of the rest of the molecule.

The O-H bond, present in alcohols, carboxylic acids and phenols, produces a broad band at around 3500 cm^{-1} . The shape and position of the band are sensitive to dilution and the nature of the solvent, due to the existence of *hydrogen* bonds which modify the energy of the O-H bond.

The C-H bond gives a thin band between 2800 cm^{-1} and 3000 cm^{-1} , but this zone provides little information, given the frequency of existence of these bonds in organic molecules. Two characteristic absorptions are the C-H bond in the CHO *formyl* group of an aldehyde, which absorbs at around 2800 cm^{-1} and the [H-C]□C bond in a terminal alkyne, at around 3300 cm $.^{-1}$

The triple bond C□N, present in nitriles, gives a fine band at around 2250 cm^{-1} . Similarly, the unsymmetrical triple bond C□C, present in alkynes, absorbs at around 2150 cm $.^{-1}$

The C=O double bond of the carbonyl group produces a thin band at around 1700 cm^{-1} for aldehydes, ketones and esters, and at around 1800 cm^{-1} and beyond for acid chlorides and anhydrides. Conjugation reduces the absorption wavelength by around 30 cm $.^{-1}$

C=C double bonds in aromatic rings and non-alkenes (vinyl group) produce thin bands at around 1600 cm $.^{-1}$

The C-O single bond present in esters, epoxides, ethers and alcohols gives a thin band between 1000 cm^{-1} and 1300 cm^{-1} , difficult to identify with certainty.

The previous table is used to assign absorptions to the various chemical groups present.

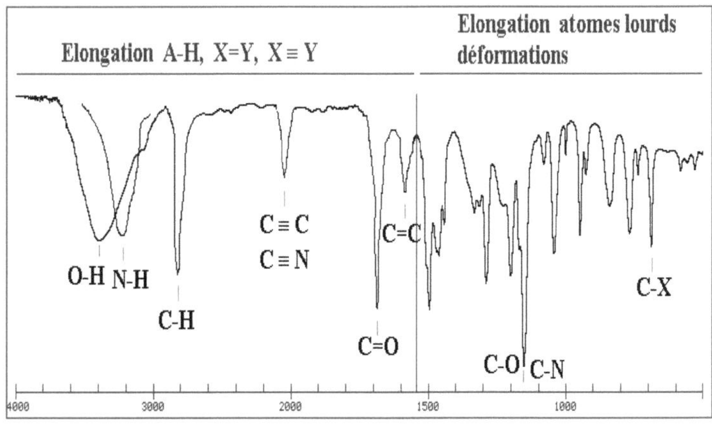

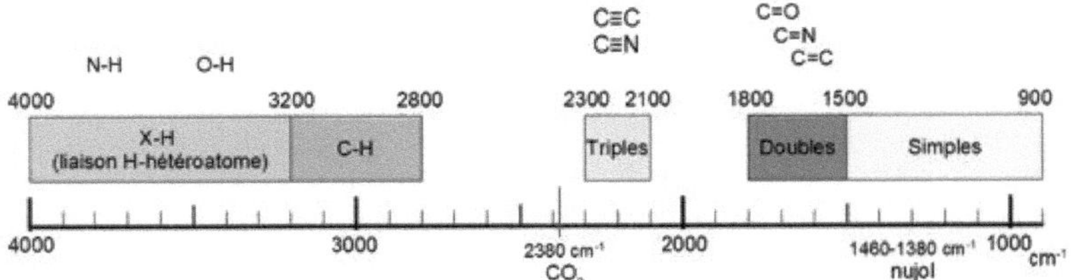

Figure 61: Infrared absorption of groups

IV. Functional analysis

➢ Determining the functional groups of a molecule: alcohol, aldehyde, ketone, acid...

➢ Determining the bonds between carbons in a chain: saturated, unsaturated, aromatic character of a molecule...

1. Alkanes

The IR spectra of alkanes show the bands of v elongation and δ deformation vibrations of C-H and C-C bonds.

δ(C-C) appear in the $\bar{v} < 500$ cm range^{-1} , they are not always observed.

v(C-C) appear in the 1200-800 cm range .$^{-1}$

Being weak, these bands do not help identification.

δ(C-H) appear in the 1475-1340 cm range^{-1} , they are of high intensity.

v(C-H) appear in the 3000- 2840 cm range^{-1} , they are of high intensity.

Table 16: Some bands of elongation v and deformation δ vibrations of C-H bonds.

	CH₃	CH₂
v_{as}	2962 cm⁻¹	2926 cm⁻¹
v_s	2872 cm⁻¹	2853 cm⁻¹
δ_{as}	1450 cm⁻¹	1465 cm⁻¹
δ_s	1375 cm⁻¹	

• Consider the IR spectrum of octane.

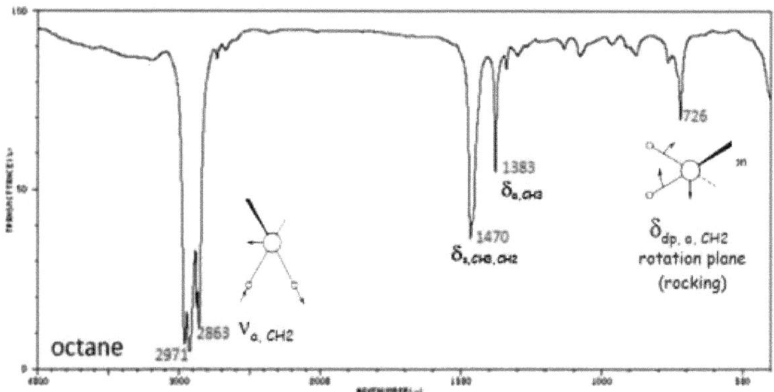

Figure 62. IR spectrum of octane.

Between 2840 cm⁻¹ and 3000 cm⁻¹ , we note the presence of C-H bond elongation vibrations:

$_{as} \square(CH_3) = 2971$ cm⁻¹ , and $v_{as} (CH_2) = 2863$ cm⁻¹

At around 1400 cm⁻¹ , these are the bands with deformation vibrations in the plane of the C-H bond:

$_{as} \square(CH_3) = 1450$ cm ,⁻¹ $_{as} \square(CH_3) = 1383$ cm ,⁻¹ $_s \square(CH_{,3} CH_2) = 1465$ cm ,⁻¹ and around 800 cm⁻¹ , $\delta_{as} (CH_2) = 726$ cm⁻¹

• Consider the IR spectrum of hexane.

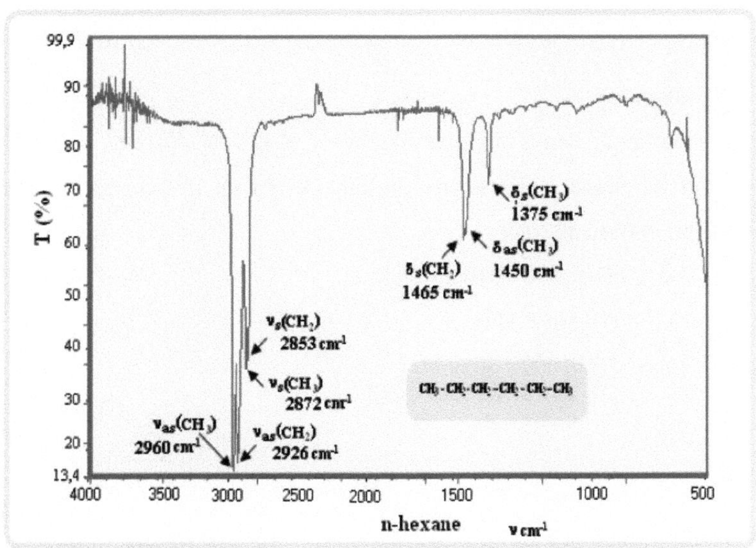

Figure 63: IR spectrum of n-hexane

Between 2840 cm^{-1} and 3000 cm^{-1} , we note the presence of C-H bond elongation vibrations:

$_{as} \square(CH_3) = 2960$ cm^{-1} $v_{as}(CH_3) = 2872$ cm^{-1}

$_{as} \square(CH_2) = 2926$ cm^{-1} $v_{as}(CH_2) = 2853$ cm^{-1}

At around 1400 cm^{-1} , these are the bands with deformation vibrations in the plane of the C-H bond:

$_{as} \square(CH_3) = 1450$ cm^{-1} , $_s \square(CH_3) = 1375$ cm^{-1} , and $\delta_s (CH_2) = 1465$ cm^{-1}

2. Cyclic alkanes

The **C-H** bond elongation vibration band, v(C-H), appears in the 3000-2990 cm range^{-1} . Increasing the tension in the cycle increases the vibrational frequency.

If there is no voltage, the elongation vibration band v(C-H) frequency appears at the same frequency for a ring and an aliphatic.

In the case of δ (C-H), cyclization decreases the vibrational frequency

For example:

	Cyclohexane	hexane
δ (C-H)	1452 cm^{-1}	1465 cm^{-1}

3. Alkenes

Three vibrational bands are possible: v (C=C), v (C-H) and δ (C-H). Vibration frequencies depend on the alkene's mode of substitution and conformation (cis and trans).

> v (C=C): 1680-1610 cm^{-1}
>
> v (C-H): 3095-3010 cm^{-1}
>
> δ (C-H) : 1000-650 cm^{-1}

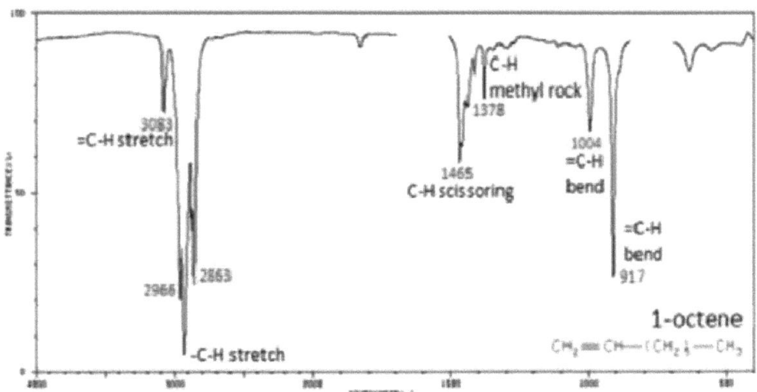

Figure 64: IR spectrum of 1-octene.

4. Alcynes

Considering the spectrum of hex-1-yne, the characteristic alkyne bands are:

ν (HC≡CH): 2100-2260 cm^{-1}

ν (C-H): 3330-3267 cm^{-1}

δ (C-H): 700-610 cm^{-1}

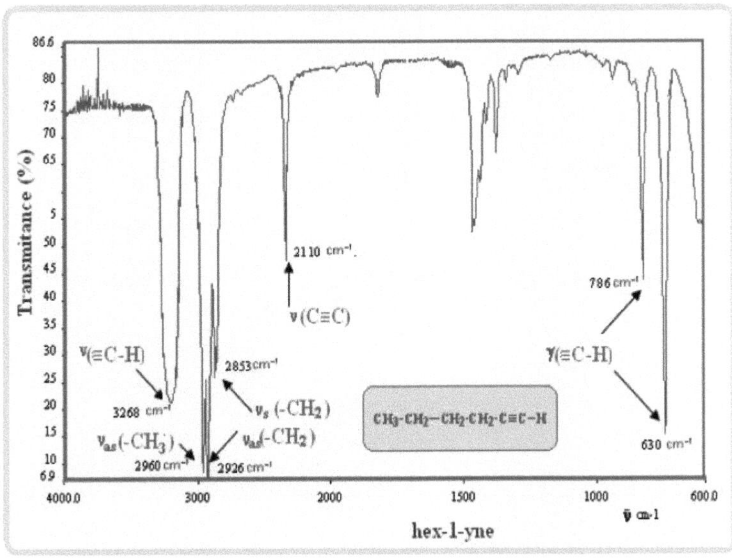

Figure 65: IR spectrum of hex-1-yne

In addition to the bands shown in the previous spectrum, there are new bands with frequencies :

Around 3268 cm^{-1} , strong elongation vibration ($\equiv$ C-H); this band appears in the case of a monosubstituted alkyne.

Around 2110 cm^{-1} , low-intensity elongation vibration (C$\equiv$ C)

Around 786 cm^{-1} and 630 cm^{-1} , deformation vibration ($\equiv$ C-H)

5. Monosubstituted aromatic compound

Aromatics have vibrational bands with which they can be easily identified:

ν(C=C): 1600-1500 cm $^{-1}$

ν(C-H): 3100-3010 cm $^{-1}$

δ(C-H): 900-690 cm^{-1}

The position of these bands depends on the substitution of the ring, and therefore on the number of neighboring Hs.

Table 17: Some vibrational bands of variously substituted aromatics

Group	δ band (C-H)	Degree of substitution
5 H neighbors	710-685 cm^{-1}	mono substitution
4 H neighbors	760-740 cm^{-1}	1.2 di substitution
3 H neighbors	800-770 cm^{-1}	1,2,3 tri substitution
		1.3 di substitution
2 H neighbors	840-800 cm^{-1}	tetra substitution
		1,3,4 tri substitution
		1.4 di substitution
1 H neighbor	900-800 cm^{-1}	

• Consider the IR spectrum of toluene:

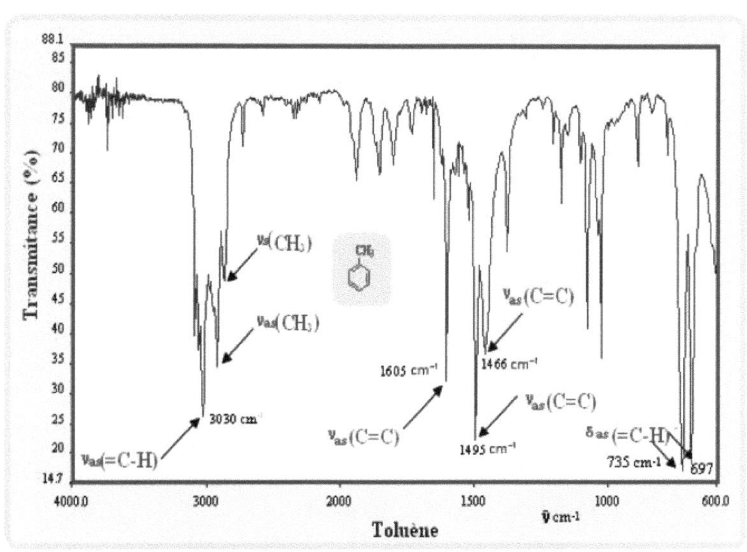

Figure 66: IR spectrum of toluene

$\nu_{as}(CH_3) = 3008$ cm⁻¹ and $\nu_s(CH_3) = 3000$ cm⁻¹

$\nu_{as}(C=C) = 1605$ cm⁻¹ , 1495 cm⁻¹ and 1466 cm⁻¹ : high-intensity bands, corresponding to elongation vibrations of the aromatic backbone.

$\nu_{as}(=C-H) = 735$ and 697 cm⁻¹ : high intensity bands, attributed to out-of-plane deformation modes of the 5 adjacent hydrogen atoms.

• **In general:**

In the case of a di-substituted benzene, three regions are of particular interest:

3030 cm⁻¹ : phenyl H elongation vibration

1500 cm⁻¹ - 2000 cm⁻¹ : elongation vibration of the aromatic skeleton

650 cm⁻¹ - 1000 cm⁻¹ : the fors deformation vibration band of the C-H bond plane is invaluable for identifying the isomers of poly-substituted aromatics.

The order of magnitude of out-of-plane vibrations of C-H bonds for di-substituted benzene rings is given.

Compound	otho	meta
	para	

81

| ν (cm)$^{-1}$ | 730 - 770 | 750 - 810 | 800 - 860 |

• **Example**: Para and meta toluene compounds

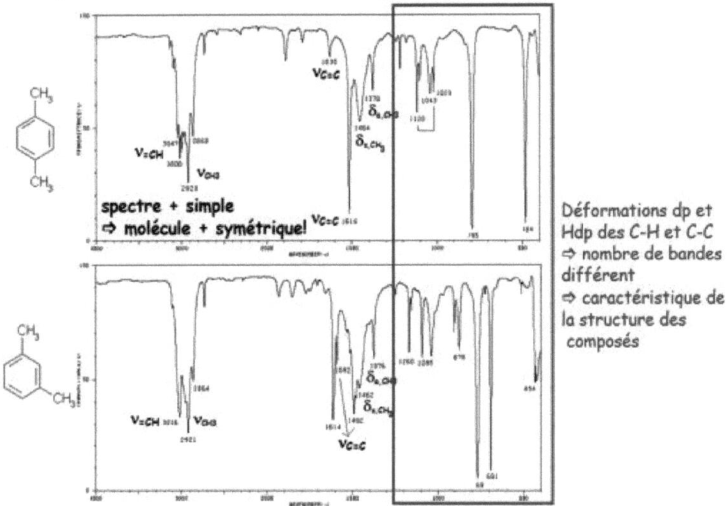

Figure 67: Structural characteristics of para- and meta-toluene compounds

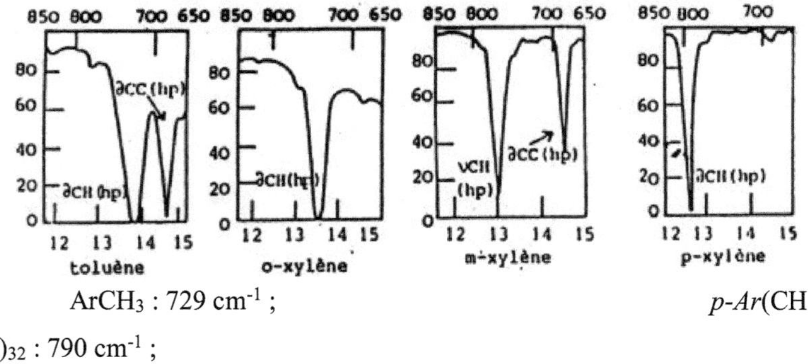

ArCH₃ : 729 cm⁻¹ ; p-Ar(CH)₃₂ : 790 cm⁻¹ ;

Figure 68: Vibrations of benzene derivatives in the 900-700 cm zone⁻¹

- ArNO₂ : 729 cm⁻¹ ; ArCOOH: 808 cm⁻¹ ; p-ArCl₂ : 880 cm⁻¹

• **Example**: Large footprint of para- and meta-toluene compounds

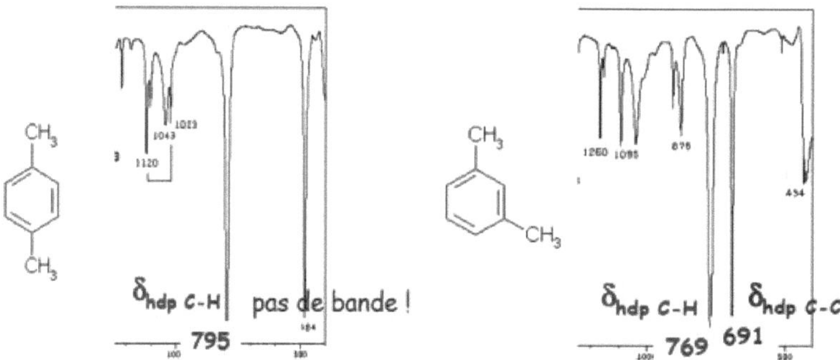

Figure 69: Fingerprint zone for para- and meta-toluene compounds

- **Example**: Large footprint of para- and meta-toluene compounds (continued)

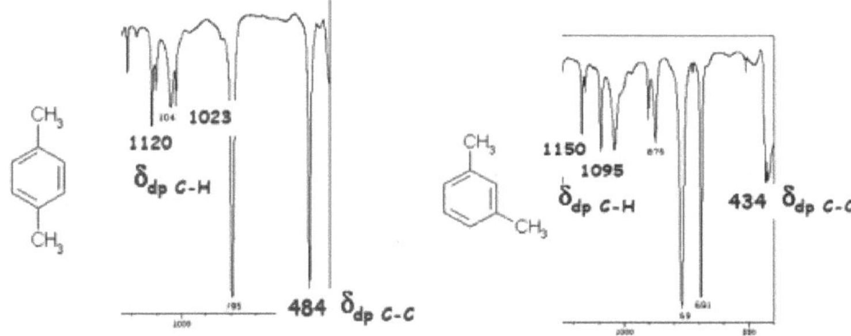

Figure 70: Fingerprint zone for para- and meta-toluene compounds

(continued)

♣♣ **Summary**

The C-H bond can be summarized in the following table:

Table 18: C-H vibrations of alkanes, alkenes, alkynes and aromatics

Link	Compound type	Vibration mode	$\bar{v}$	Intensity
C-H	Alkanes	v	2970-2850 cm^{-1}	Strong
		δ	1470-1340 cm^{-1}	Strong
C-H	Alkenes	v	3095-3010 cm^{-1}	Average
		δ		Strong

			995-675 cm^{-1}	
C-H	Alcynes	ν	3330-3267 cm^{-1}	Strong
		δ	700-610 cm^{-1}	Strong
C-H	aromatic	ν	3100-3010 cm^{-1}	Average
		δ	900-690 cm^{-1}	Strong

6. Alcohols and phenols

Characteristic bands arise from elongations ν(O-H) and ν(C-O) and deformation δ (O-H).

Elongation vibration ν (O-H): these vibrational bands are very broad and characteristic of the alcohol function.

Free OH absorbs intensely between 3700-3584 cm^{-1}.

OHs are often involved in hydrogen-bridge bonds, which affects the vibrational frequency of ν (O-H).

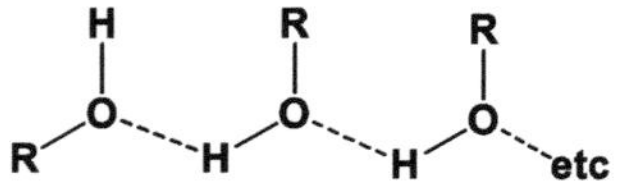

Liaison H intermoléculaire

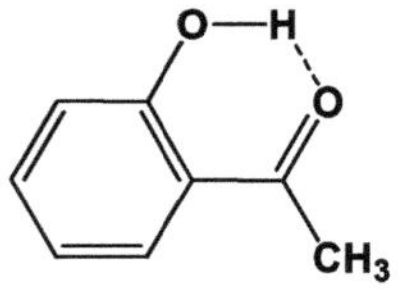

Liaison H intramoléculaire

Figure 71: Diagram of intermolecular and intramolecular hydrogen bonding

Elongation vibration ν (C-O): 1260-1000 cm^{-1}.

Strain vibration δ (O-H): in-plane:1420-1330cm^{-1} and out-of-plane: 769-650cm^{-1}

The characteristic bands of the C-O and O-H bonds, depending on the type of alcohol, are summarized in the following table:

Table 19: Vibrations of alcohols (I, II and III) and phenol

	Alcohol I	**Alcohol II**	**Alcohol III**	**Phenol**
$_{O-H}\ \square(cm\ )^{-1}$	1208	1355	Around 1380	1360
$_{O-H}\ \square(cm\ )^{-1}$	1017	1138	Around 1160	1223

- Consider the IR spectrum of ethanol:

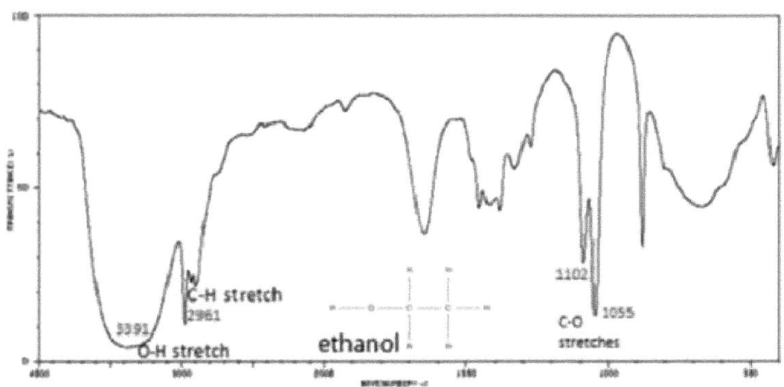

Figure 72: IR spectrum of ethanol

- Consider the IR spectrum of butanol:

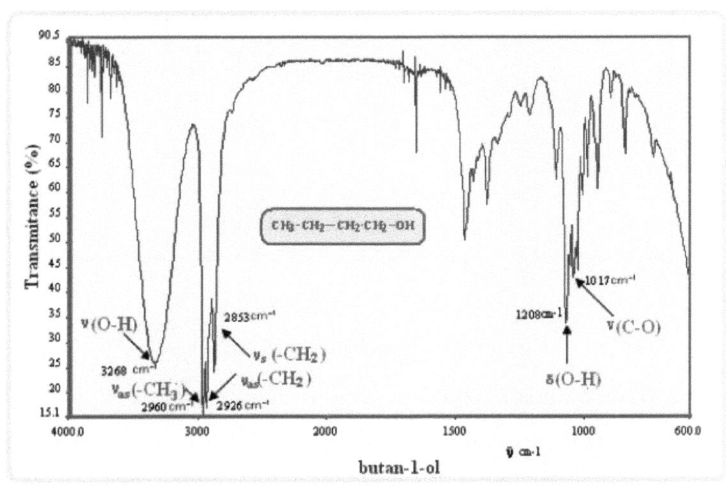

Figure 73: IR spectrum of butanol

For an alcohol, we find a wide band located between 3200 cm⁻¹ and 3400 cm . ⁻¹

The spectrum is essentially :

A wide band around 3268 cm^{-1} attributed to the O-H symmetrical elongation vibration.

A deformation vibration at around 1208 cm^{-1} corresponds to the O-H bond.

An asymmetrical elongation vibration around 1017 cm^{-1} corresponding to the O-H bond.

7. Ether

The characteristic response of ethers is associated with the elongation of the C-O-C system. There is a symmetrical elongation band around 1030 cm^{-1} and an asymmetrical elongation band that is always strong around 1200 cm^{-1}.

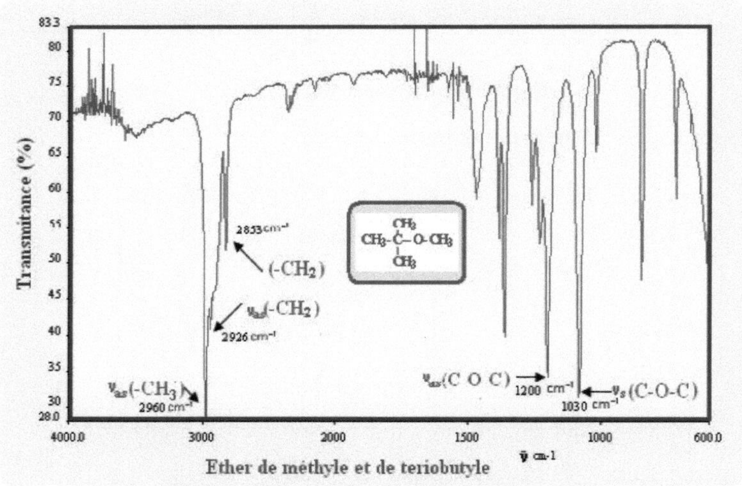

Figure 74: IR spectrum of an ether

8. Ketones and aldehydes

The carbonyl function is one of the functions that can be characterized very easily by infrared, by the elongation vibration C=O, which presents a very intense and fine band between 1870-1540 cm^{-1}.

They are also characterized by the elongation and deformation vibration of C-CO-C in the 1300-1100 cm range.$^{-1}$

In the case of aldehydes, we also have the elongation vibration ν (C-H): 2900-2695 cm^{-1}

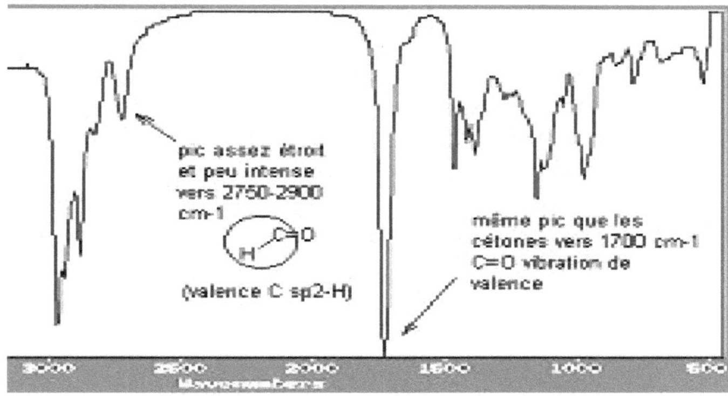

Figure 75: IR spectrum of an aldehyde

♣ IR spectrum of butanone

All organic compounds with a carbonyl group have an intense band at around 1700 cm^{-1} . This is the most intense and sharpest band in IR.

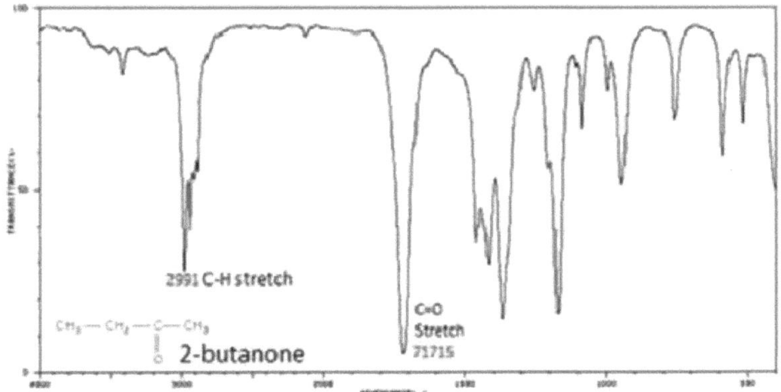

Figure 76: IR spectrum of butan-2-one

♣ IR spectrum of butanal

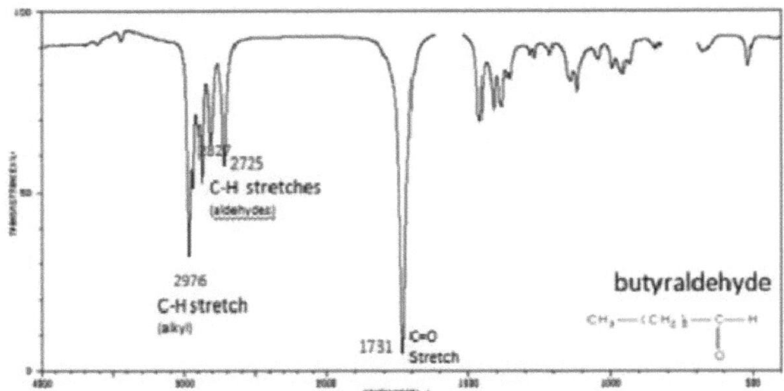

Figure 77: Butanal spectrum

♣ IR spectrum of 2-methylpropanal

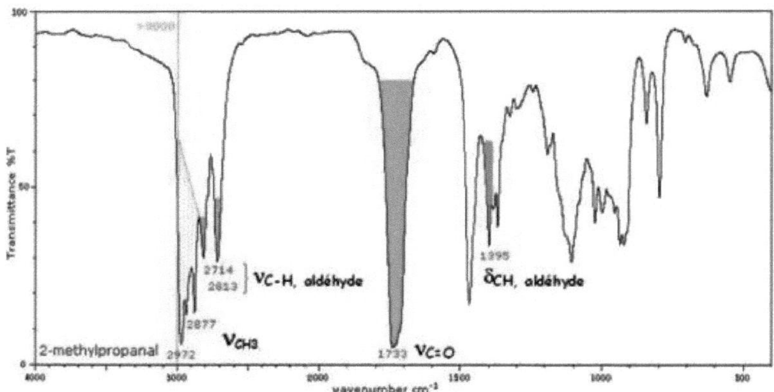

Figure 78: IR spectrum of 2-methylpropanal

9. Carboxylic acid

In solution or in the solid state, carboxylic acids exist as dimers due to strong hydrogen bonds.

Figure 79: Diagram of carboxylic acid dimer (strong hydrogen bonds)

The elongation vibration band of the free O-H bond only appears for very dilute solutions at around 3520 cm .$^{-1}$

In the case of dimers, this band appears in the 3300-2500 cm range[-1] due to strong hydrogen bonds.

The ν (C=O) vibration band appears in the 1760-1700 cm range[-1], more intense than those of aldehydes and ketones, its position depending on the presence of H bonds.

Two other bands are characteristic of carboxylic acids, that of the elongation vibration ν (C-O), which appears in the 1320-1210 cm range[-1], and that of the deformation δ (O-H) in the 1440-1395 cm area.[-1]

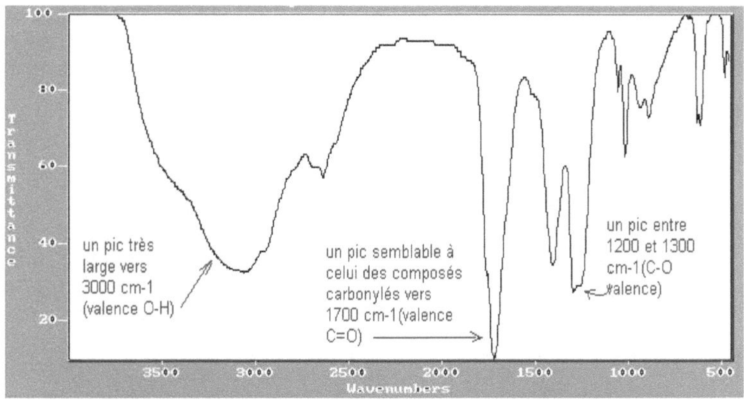

Figure 80: IR spectrum of a carboxylic acid

♣ IR spectrum of hexanoic acid

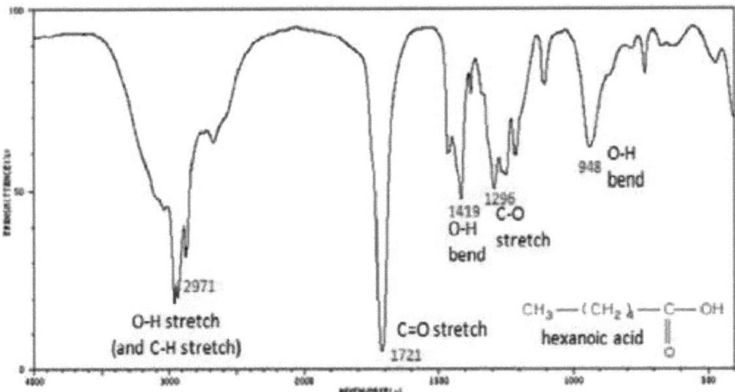

Figure 81: IR spectrum of hexanoic acid.

10. Amines

Primary amines -NH₂ exhibit two ν (N -H) vibrational bands, one in the 3400-3300 cm range⁻¹ and the other in the 3330-3250 cm range .⁻¹

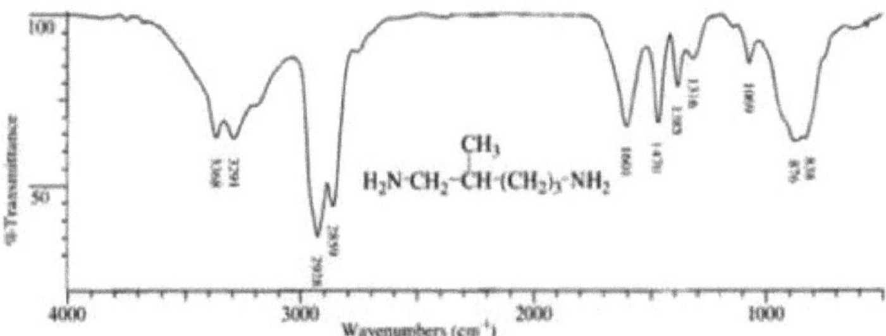

Figure 82: IR spectrum of 2-methylpentan-1,5-diamine.

Secondary amines exhibit a single ν (N-H) vibrational band in the 3350-3310 cm range . ⁻¹

These bands are thinner than those of ν (O-H).

Amines are also characterized by the presence of the following bands:

The medium-to-strong δ (N-H) deformation band in the range 1650-1580cm⁻¹ for primary amines, and around 1515 cm⁻¹ in the case of secondary amines.

The elongation band ν (C-N) of low to medium intensity in the range 1250-1020cm^{-1} for aliphatic amines, and in the range 1342-1266 cm^{-1} for aromatic amines.

11. Nitriles

Nitriles show a characteristic band of the C≡N bond in the range 2280-2210 cm.$^{-1}$

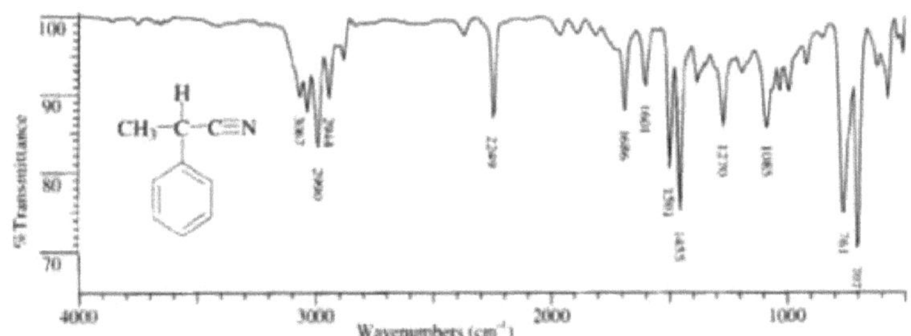

Figure 83: IR spectrum of 2-phenylpropanenitrile.

V. Structural study

1. Parameters influencing group frequencies

$$\overline{\nu}\,[cm^{-1}] = \frac{1}{2\pi c}\sqrt{\frac{k}{\mu}}$$

The binding force k of the vibrator, and therefore its vibration frequency, can be influenced by the "effects" of neighboring vibrators.

91

Figure 84: Diagram of parameters influencing group frequencies

1. Attractive inductive effect (-I) of chlorine

2. Conjugation with the aromatic ring

3. Steric hindrance

4. Highly stable intramolecular hydrogen bond (6-ring)

5. Mechanical effect of the cycle to stiffen the vibrator C=O

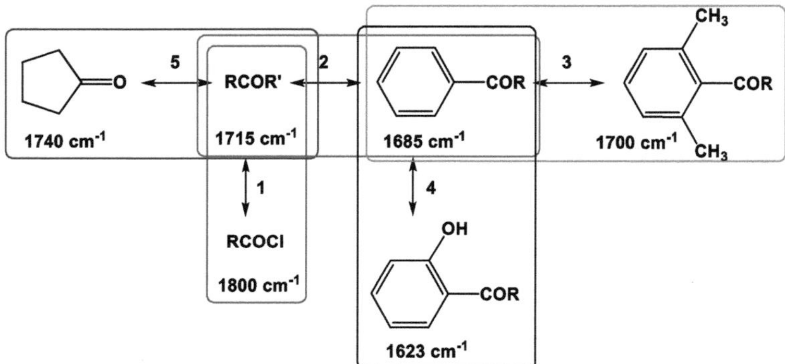

Figure 85: Diagram of parameters influencing group frequencies (continued)

1.1. Hydrogen bonding

The X-H bond, where X is a heteroatom (O, N, S), can be involved in molecular associations of the hydrogen bond type.

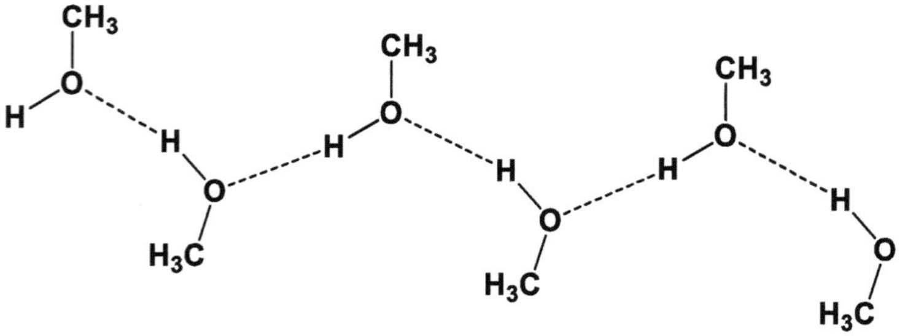

Figure 90: Hydrogen bonding in pure methanol

<u>Effects of hydrogen bonding :</u>

The effect of hydrogen bonding is :

➢ Weakening of the X-H bond (reduction in the number of vibrational waves)

➢ To cause a broadening of the corresponding band due to the vibration ν_{XH}.

1.1.1. Intermolecular hydrogen bonding in alcohols:

⁂ Consider the IR spectrum of hexan-1-ol in the pure liquid state (1), in dilute solution in CCl_4 as solvent (2) and in the gaseous state (3).

Bandwidth: intermolecular interactions

Bandwidth depends on number of vibrator chemical environments >>> The strength of intermolecular bonds, such as the hydrogen bond, influences the bond strength k of the vibrator, and therefore its vibration frequency.

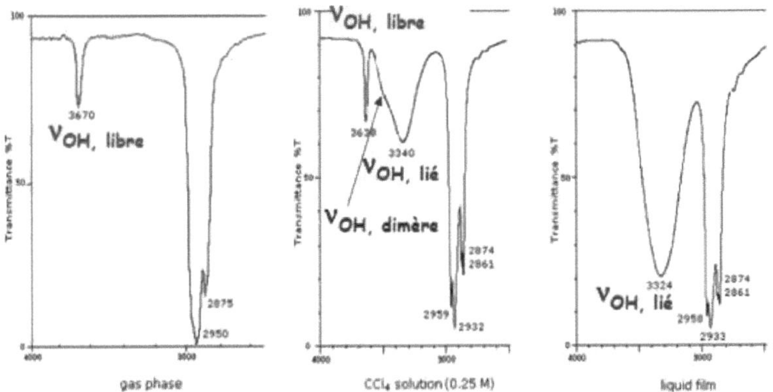

Figure 91: Spectrum of hexan-1-ol in the pure liquid state, in the gaseous state and diluted in CCl_4

Experimental conditions!

T°, [ROH], solvent

Pure product

The wide band between 3200 cm⁻¹ and 3400 cm⁻¹ is attributable to the OH valence vibration of an OH vibrator engaged in hydrogen bonding. >>> OH associated by hydrogen bonding: $\nu_{OH\ associated\ (bonded)}$.

<u>**Dilution in an aprotic solvent such as CCl$_4$**</u>

By dilution in an aprotic solvent such as CCl$_4$, this broad band disappears in favor of the appearance of a thin band located in the 3590-3650 cm zone^{-1} : free vOH.

This behavior shows that the nature of the hydrogen bond in the alcohol studied is intermolecular.

<u>**For example:**</u>

(1) Hexan-1-ol in the pure liquid state (Figure 92)

(2) Hexan-1-ol solution diluted in CCl$_4$ (Figure 92)

→ Hydrogen bonding in the alcohol studied is intermolecular **O—H----O**

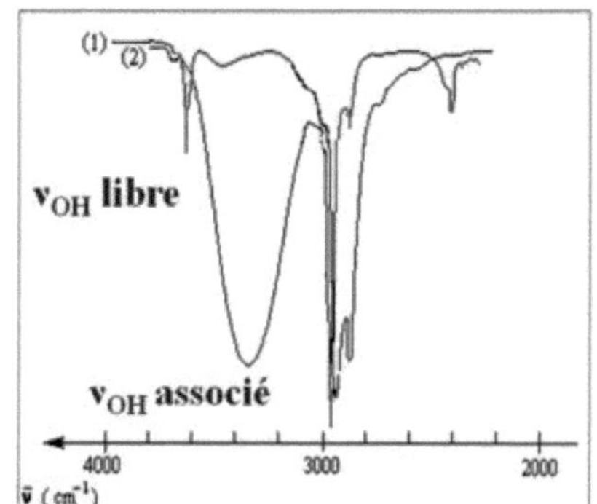

Figure 92: Free and associated v_{OH} band

1.1.2. Intermolecular hydrogen bonding in acids

➢ The IR spectra of carboxylic acids show a much wider vOH band than for alcohols, and at a lower frequency, giving the spectrum a highly characteristic appearance.

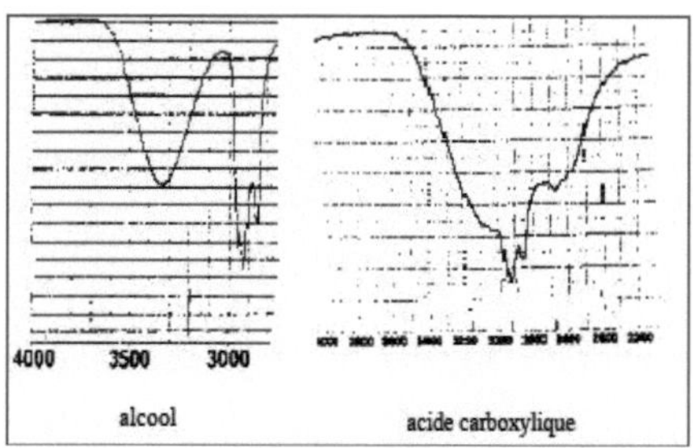

Figure 93: ν-band$_{OH}$ of a carboxylic acid and an alcohol

➢ Carboxylic acids exist as dimers due to the very strong hydrogen bonds between O-H and C=O:

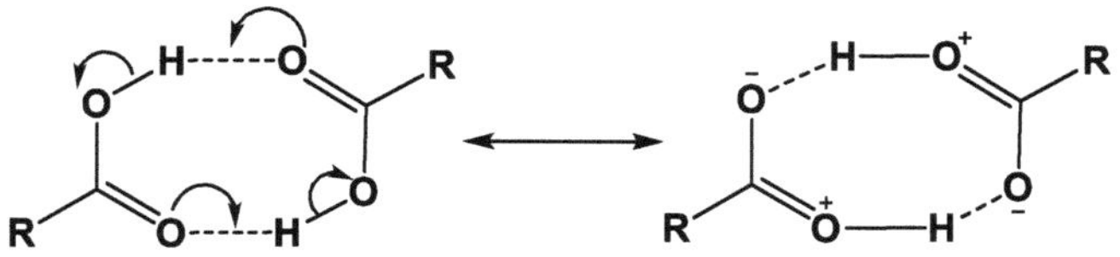

Figure 94: Hydrogen bonding diagram

➢ The absorption band ν_{OH} , very wide and very intense, is located between 3300 and 2500 cm^{-1} . The bands due to ν_{CH} are superimposed on this band. In very dilute solution in a non-polar solvent, ν_{OH} moves towards 3520 cm^{-1} (monomeric form, free OH).

➢ For the H-bond acceptor carbonyl, in the dimer, the C=O bond is weakened by the H bond. Its frequency $\nu_{C=O}$ of the monomer located around 1760 cm^{-1} and $\nu_{C=O}$ decreases by 40 to 60 cm^{-1} in the dimer compared to the ν band$_{C=O}$ of the monomer.

1.1.3. Intramolecular hydrogen bonding

Example: Polyols

In some molecules, such as polyols, intramolecular hydrogen bonds can be observed. It's easy to distinguish between intermolecular and intramolecular bonds using infrared spectroscopy. By dilution in a solvent such as CCl_4 , the absorption band due to the former is shifted, while that due to the latter remains unchanged.

1.2 Inductive and mesomeric effects

Let's take the case of a carbonyl function. The valence vibration $\nu_{C=O}$ for an aliphatic ketone is around 1715 cm . $^{-1}$

⁂ Inductive attractive effects↗ : $\nu_{C=O}$ ↗

Attractive inductive effects tend to increase this frequency.

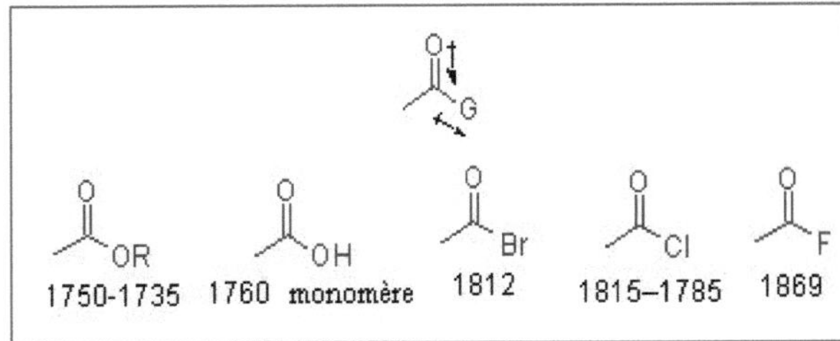

Figure 95: Attractive inductive effects on carbonyl vibration frequencies

♣ Mesomeric effects (resonance)↗ : $\nu_{C=O}$ ↘

On the other hand, resonance-inducing mesomeric effects lead to a decrease in frequency $\nu_{.C=O}$

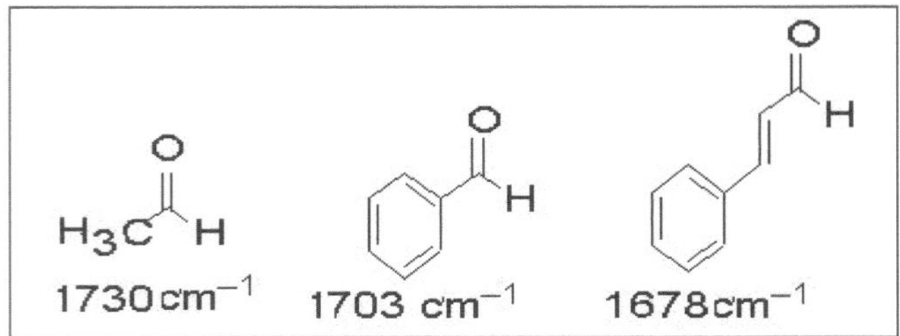

Figure 96: Mesomeric effects on carbonyl vibration frequencies

1.3.. Conjugation

The delocalization of a double bond reduces its force constant, and therefore its vibrational frequency.

<u>Examples</u>:

Carbonyl conjugate, $\nu_{C=O}$ decreases from 15 to 40 cm^{-1}

Figure 97: Vibration frequencies of conjugated carbonyls

1.4. Cycle voltage

When the oscillator is connected to a sterically tensioned structure, its vibration frequency is increased.

Examples:

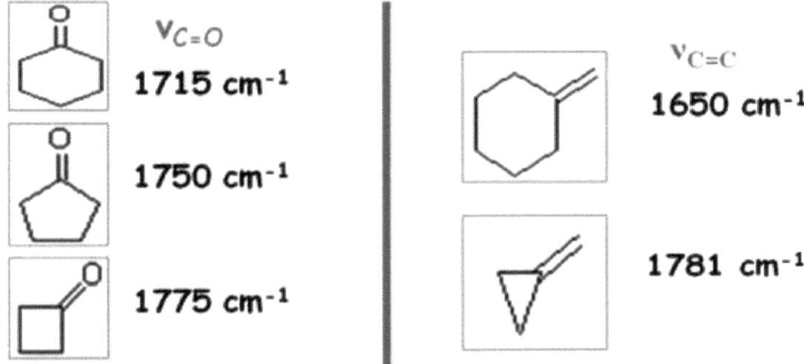

Figure 98: Vibration frequencies of ring-bound C=O and C=C

1.5. Isomers

IR spectrometry is used to differentiate isomers.

Examples: cis and trans isomers of olefins.

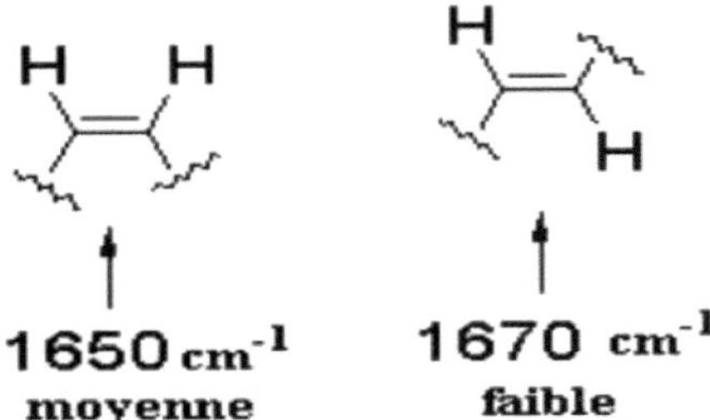

Figure 99: Vibration frequencies of cis and trans olefin isomers

VI. Equipment

There are two types of IR spectrometer:

A scanning IR spectrometer: the most classic model, similar to the spectrophotometers used in UV-visible spectroscopy.

A Fourier Transform IR spectrometer (FTIR) is identical to a scanning spectrometer, with the dispersive system replaced by a Michelon interferometer whose position is adjusted by laser.

They consist of the following components:

Source

Sample

Dispersive system

Detector

Generally speaking, the sources and detectors for both types of spectrometer can be the same. Schematically, the device looks like this:

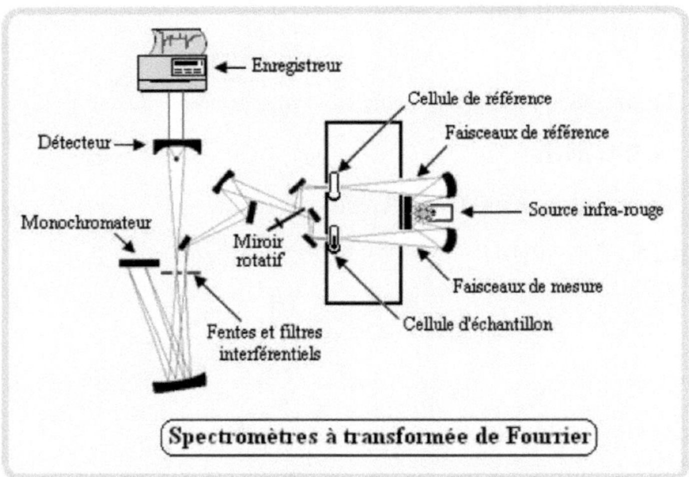

Figure 100: Fourier transform IR spectrometer (FTIR)

1. How an IR spectrometer works

The choice of source depends on the infrared region you wish to work in. In most cases, however, we work in the mid-infrared region (4000 and 400 cm^{-1}). We generally use a Globar source based on silicon carbide.

The IR radiation coming from the source is polychromatic, first split into two equivalent beams, one of which is focused on a reference cell and the other on a cell containing the sample. The latter then passes through the sample compartment and, thanks to a rotating sector mirror, is recombined with the reference beam. This recombined beam then passes through the grating monochromator slit.

The monochromator usually consists of a grating capable of dispersing the incident radiation into its various wavelengths. This grating is also in constant rotation, so that each wavelength can be focused one after the other on a detector.

The most widely used detector is a pyroelectric detector. This is a Deuterium Tryglycine Sulfate (DTGS) crystal. Below a temperature known as the "Curie point", ferroelectric bodies such as DTGS exhibit strong spontaneous polarization between certain faces of the crystal. If the temperature of such a crystal changes, under the action of IR radiation, its polarization changes. The result is a variation in voltage as a function of the variation in crystal temperature. This detector detects temperature variations and transforms them into intensity variations. This difference in intensity is then easily converted into transmittance (T), usually expressed in %.

The intensity of the beam arriving at the detector is translated into an interferogram, which is then processed by Fourier Transform. This is a mathematical process that breaks down a complex, time-dependent signal into a sum of simple signals of known frequencies, according to the relationship :

$$I'(x) = \frac{I(\gamma)}{2}(1 + \cos(2\pi\lambda x))$$

I ($\square$) is the radiation intensity supplied by the source

$I'(x) = \frac{I(\gamma)}{2}(\cos(2\pi\lambda x))$ is the interferogram

I (x) can be calculated by : $I(\gamma) = \int_{-\infty}^{+\infty} I(x)\cos(2\pi\gamma x)dx$

Interferograms are acquired and transformed into spectra by a mini-computer built into the spectrometer.

2. Sample preparation

IR spectra of solid, liquid and gaseous compounds can be produced. Depending on the sample, KBr-based pellets are used, either in cuvettes, or a liquid is deposited between the blades of the KBr pellets.

For solids: pellets are generally prepared from a mixture of sample (1%) in powder form mixed with KBr, which is transparent to IR radiation. The mixture is finely ground and mixed in a mortar.

Liquids: low-viscosity or volatile liquids are introduced into a closed vessel of specified thickness. Viscous and low-volatility liquids are deposited between KBr plates.

Gases are introduced into a tank with a larger volume than that used for liquids.

Example of IR monitoring of an organic reaction:

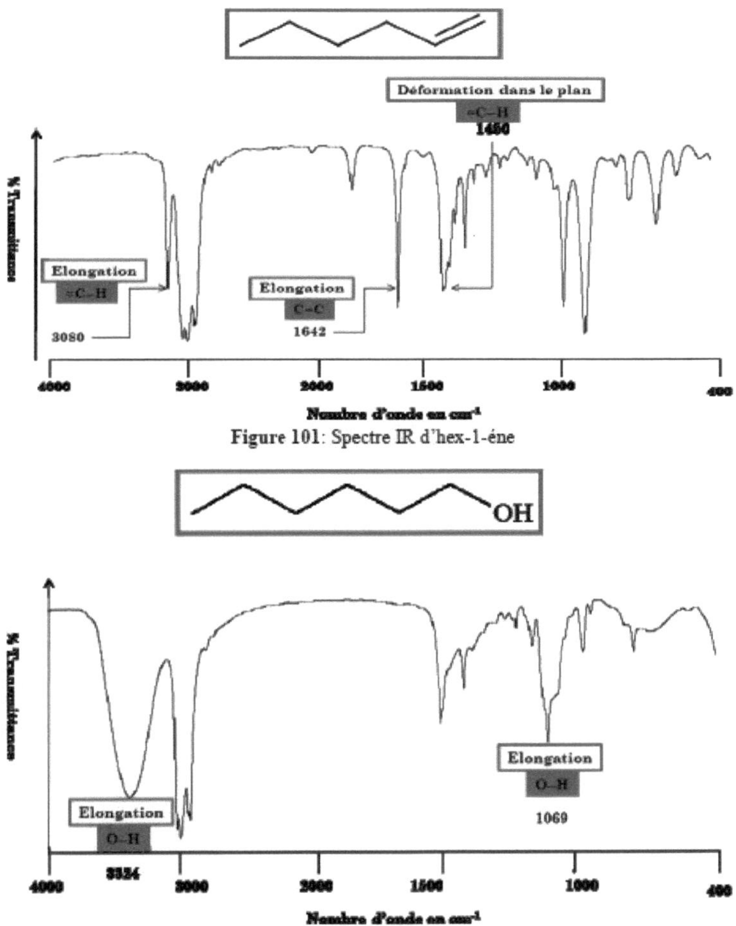

Figure 101: Spectre IR d'hex-1-éne

Figure 102: IR spectrum of hexan-1-ol

VII. Interpreting an IR absorption spectrum

Analysis of band allocation provides two levels of information:

The nature of the functions present in the molecule

The structural nature

But complete identification of the molecule is very rare. Bands are analyzed according to their :

Position $(cm)^{-1}$

Intensity (low, medium, high)

Shape (wide or narrow)

IR spectrometry therefore provides a rapid solution foridentifying an organic compound. All that's needed is to check the identity of each of the bands in the reference spectrum and the spectrum of the product under study, traces under the same sampling conditions (gas, liquid or solid) and with equipment of the same performance. The wavelength should be considered as determined with a precision of 5 to 10 cm^{-1} depending on the areas of the spectrum.

1. Characteristic elongation and deformation frequencies of atomic groupings in IR

The ranges attributed to various elongation and deformation frequencies, as well as the characteristics of some groups of atoms, are shown in Table 14 of I n f r a r e d vibrations... although these ranges are relatively narrow, it should be borne in mind that the exact frequency at which a given group absorbs may vary according to the environment existing within the molecule and the physical state of the substance.

Table 20: Elongation and deformation frequencies of selected groups of atoms.

Group	Link	Wave number (cm)$^{-1}$	Vibration	Intensity
Alcohols and phenols	Free O-H	3650-3590	elongation	variable and fine
Alcohols and phenols	O-H assoc.	3400-3200	elongation	strong and broad
Acids	O-H assoc.	3300-2500	elongation	strong and wide-ranging
Primary amines	N-H	3500 / 3410	asymmetrical elongation / symmetrical elongation	average / average
Secondary amines	N-H	3500-3310	elongation	average
≡C-H (alkynes)	C-H	~3300	elongation	medium and fine

Aromatics	C-H	3080-3030	elongation	variable
HC=CH$_2$ (vinyl)	C-H	3095-3075 3040-3010	elongation elongation	average average
=CH$_2$ (geminal di-substituted alkenes)	C-H	3095-3075 3040-3010	elongation elongation	average average
HC=CH or C=CH	C-H	3040-3010	elongation	average
CH$_3$ (alkanes)	C-H	~ 2960 ~ 2870	asymmetrical elongation symmetrical elongation	strong strong
CH$_2$ - (alkanes)	C-H	~ 2925 ~ 2850	asymmetrical elongation symmetrical elongation	strong medium to high
-CH2-	C-H	1470	deformation	Average
-C-H (aliphatic)	C-H	2890-2880	elongation	low
-CH	C-H	1340	deformation	low
Aldehydes	C-H	2900-2800 2775-2700	elongation elongation	low average
Nitriles	C≡N	2260-2210	elongation	medium to high
Alcynes	C≡C	2140-2100	elongation	low
Aliphatic aldehydes	C=O	1740-1720	elongation	strong
Aromatic aldehydes	C=O	1715-1690	elongation	strong
Aliphatic ketones	C=O	1725-1705	elongation	strong
Aromatic ketones	C=O	1700-1670	elongation	strong
Acids	C=O	1725-1700	elongation	strong
Aliphatic esters	C=O	1750-1730	elongation	strong
Alkenes	C=C	1675-1645	elongation	average
Aromatics	C=C	1600 ; 1580 1500 ; 1450	elongation; 4 bands	variables
Nitro group (aliphatic)	C-NO$_2$	1570-1550 1380-1370	elongation elongation; 2 bands	intense

Nitro group (aromatic)	C-NO$_2$	1570-1500 1370-1300	elongation elongation; 2 bands	intense
Aliphatic amines	C-N	1220-1020	elongation	average
Aromatic amines	C-N	1360-1180	elongation	medium to high
Esters	C-O	1300-1050	elongation; 2 bands	strong
Acids	C-O	1300-1200	elongation	strong
Tertiary alcohols	C-O	1200-1125	elongation	variable
Secondary alcohols	C-O	1125-1085	elongation	variable
Primary alcohols	C-O	1085-1050	elongation	variable
Ethers	C-O	1150-1020	elongation	strong

2. IR spectrum analysis procedure

To analyze the spectrum of an unknown substance, we first identify the existence or non-existence of major functional groups.

The C=O; O-H; N-H; C-O; C=C; C=N and NO bands$_2$ are the most remarkable and, if present, immediately provide structural information.

We don't n e e d t o make a detailed analysis of C-H adsorptions at around 3000 cm-^{1} (almost all compounds have them!).

Is a carbonyl group present?

C=O groups give intense absorption in the 1820- 1600 cm region^{-1} . The band is often the most <u>intense </u>in the spectrum. Its width is medium. You can't miss it.

a - <u>If a C=O group is present </u>:

1. Acids (COOH): if OH is also present. The OH band is <u>very wide </u>3400-2400 cm^{-1} (it covers the C-H).

2. Amides (O=C-NHR) monosubstituted on nitrogen or (O=C-NH$_2$) unsubstituted on nitrogen: check that there are one or 2 N-H bands in the 3500 cm zone^{-1} (<u>medium to strong intensity</u>).

3. Esters (O=C-O-R): check for the presence of C-O (<u>intense </u>absorption around 1000 to 1300 cm).$^{-1}$

4. Anhydrides (O=C-O-C=O): if 2 C=O absorptions around 1760 and 1810 cm $.^{-1}$

5. Aldehydes (O=C-H): check for the presence of C-H aldehyde: 2 bands of medium intensity around 2750 and 2850 cm $.^{-1}$

6. Ketones (O=C-R): if the previous 5 choices are eliminated.

b - If a C=O group is absent :

1. Alcohols (-C-OH) and Phenols: look for a broad O-H band in the 3300-3600 cm zone^{-1} ; confirmed by the presence of a C-O band between 1000 and 1300 cm $.^{-1}$

2. Primary (RNH_2) and secondary (RNHR') amines: check the N-H absorption (average intensity) at around 3500 cm $.^{-1}$

- Ethers (C-O-C): look for the presence of the C-O band (and the absence of O-H) at around 1000 to 1300 cm $.^{-1}$

3. Double bond and/or aromatic ring

Alkenes: fairly weak absorption around   1650cm^{-1} . The existence of C=C is confirmed by consulting the =C-H region above 3000cm $.^{-1}$

Aromatic: medium to high intensity absorptions in the 1450 to 1650 cm region^{-1} . The existence of C=C is confirmed by consulting the =C-H region above 3000 cm $.^{-1}$

4. Triple bond

Nitriles (C≡N): medium intensity absorption, very fine peak around 2250 cm $.^{-1}$

Alkynes (-C≡C-): fine absorption of low intensity around 2150 cm^{-1} . For true alkynes, also look for C-H around 3300 cm^{-1} : fairly intense, fairly fine band.

5. NITRO" function: NO_2 has 2 strong bands around 1500 to 1600 cm^{-1} and 1300 to 1390 cm $.^{-1}$

6. Saturated hydrocarbons

If none of the above information has been observed.

If the main absorptions are located in the C-H region below 3000 cm $.^{-1}$

If the <u>spectrum is very simple</u>, only absorptions around 1450, 1375 and perhaps 720 cm $.^{-1}$

Appendices: IR spectroscopic data tables

APPENDIX 1: SIMPLIFIED TABLE OF IR SPECTROSCOPIC DATA

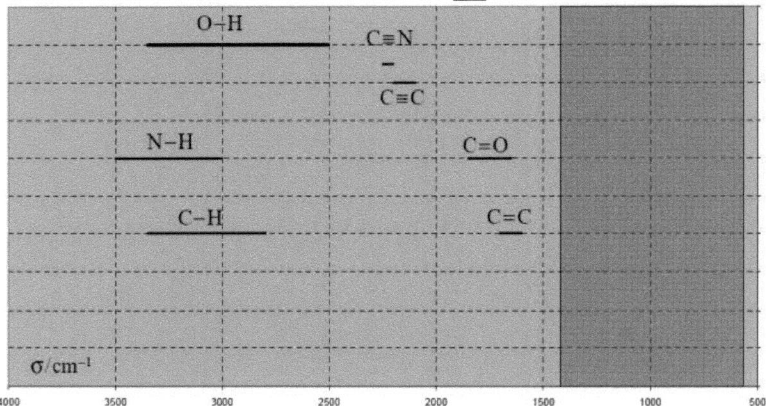

APPENDIX 2: Characteristic groups and infrared (IR) absorption bands

Fonction	Alcool	Aldéhyde	Cétone	Acide carboxylique	Alcène	Ester	Amine	Amide
Groupe caractéristique	$-O-H$ Hydroxyle	$-C\langle^O_H$ Carbonyle	$C-C\langle^O_C$ Carbonyle	$-C\langle^O_{OH}$ Carboxyle	$\rangle C=C\langle$ Alcène	$-C\langle^O_{O-C}$ Ester	$\rangle N-$ Amine	$-C\langle^O_{N-}$ Amide

Liaison	Nombre d'ondes σ (cm⁻¹)	Intensité[1]	Liaison	Nombre d'ondes σ (cm⁻¹)	Intensité[1]
$O-H_{libre}$ [2]	3580-3650	F ; fine	$C=O_{ester}$	1700-1740	F
$O-H_{lié}$ [2]	3200-3400	F ; large	$C=O_{aldéh. cétone}$	1650-1730	F
$N-H$	3100-3500	M	$C=O_{acide}$	1680-1710	F
$C_{tri}H$ [3]	3000-3100	M	$C=C$	1625-1685	M
$C_{tri}H_{aromat.}$ [4]	3030-3080	M	$C=C_{aromat.}$	1450-1600	M
$C_{tét}H$ [5]	2800-3000	F	$C_{tét}H$	1415-1470	F
$C_{tri}H_{aldéhyde}$	2750-2900	M	$C_{tét}O$	1050-1450	F
$O-H_{acide\ carb.}$	2500-3200	F ; large	$C_{tét}C_{tét}$	1000-1250	F

(1) L'intensité traduit l'importance de l'absorption : **F** : forte ; **M** : moyenne.

(2) **O–H**$_{libre}$: sans liaison hydrogène ; **O–H**$_{lié}$: avec liaison hydrogène.

(3) **C**$_{tri}$: correspond à un carbone trigonal (engagé dans une double liaison).

(4) aromat. : désigne un composé avec un cycle aromatique comme le benzène ⬡ ou ses dérivés.

(5) **C**$_{tét}$: correspond à un carbone tétragonal (engagé dans quatre liaisons simples).

APPENDIX 3: IR SPECTROSCOPIC DATA TABLE

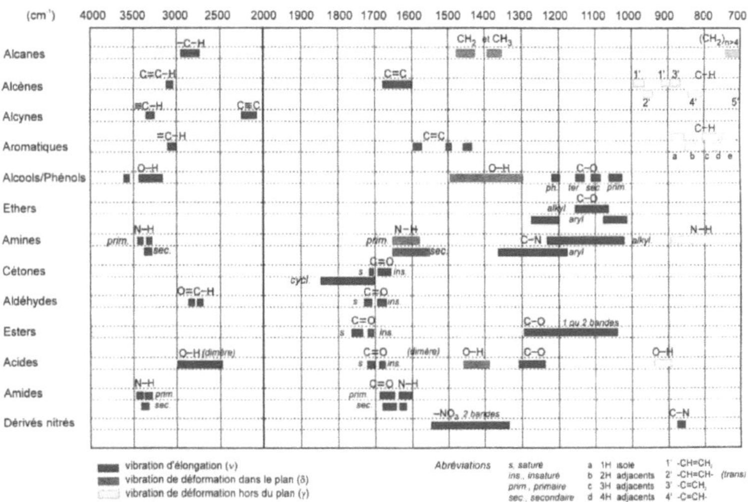

APPENDIX 4: Table of valence and strain vibration wave numbers (1)

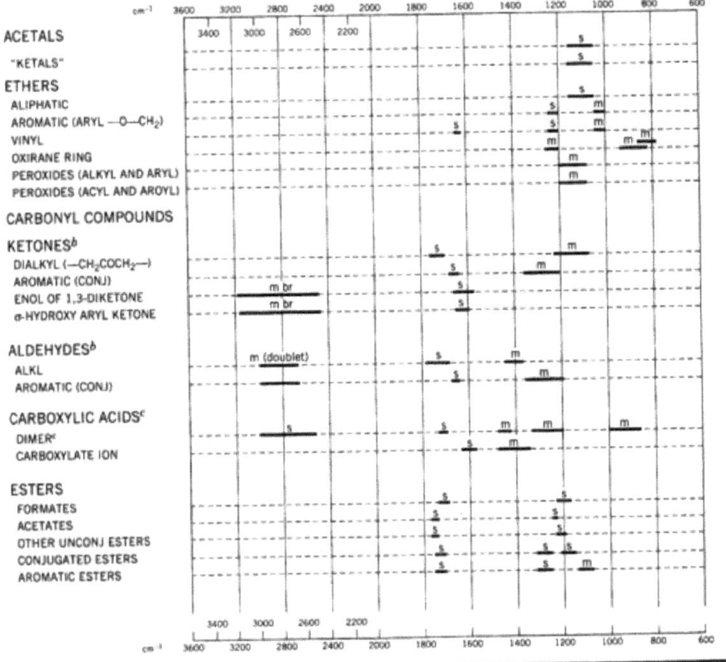

APPENDIX 5: Table of valence and strain vibration wave numbers (2)

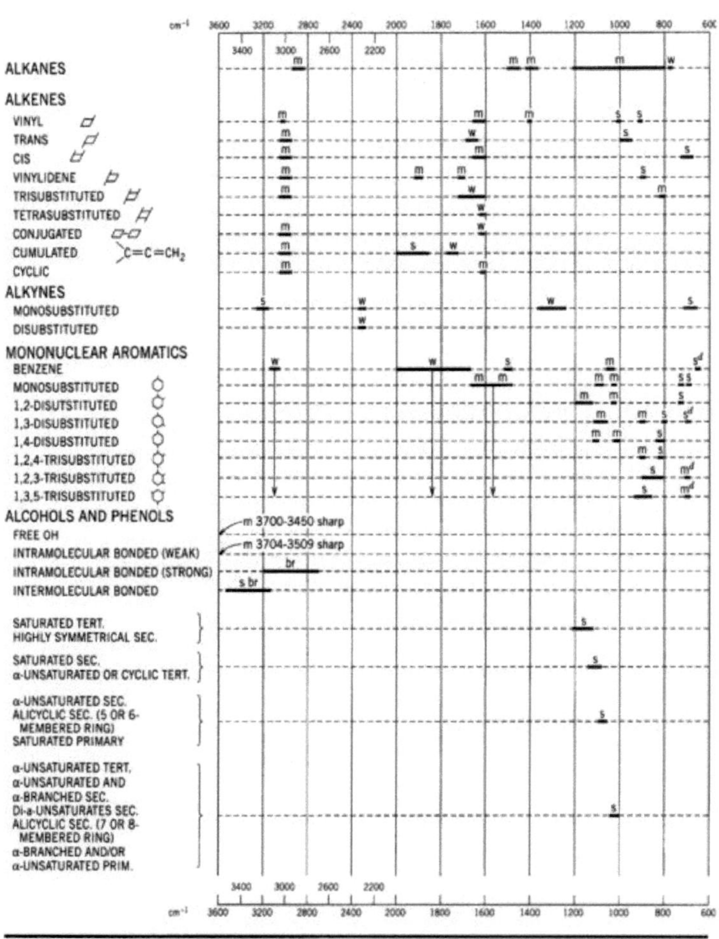

* Absorptions are shown by heavy bars. s = strong, m = medium, w = weak, sh = sharp, br = broad. Two intensity designations over a single bar indicate that two peaks may be present.

b May be absent.

c Frequently a doublet.

d Ring bending bands.

Exercises

Exercise 1:

Aniline absorbs at a wavelength $\lambda_{max} = 280$ nm with $\varepsilon_{max} = 1430$ L.mol^{-1} .cm^{-1} . We want to prepare an aqueous aniline solution with a % transmission equal to 30, when using a 10 mm thick cell. What is the mass of aniline required to prepare 100 ml of such a solution?

Exercise 2:

1) Calculate the ε_{max} of a compound with a maximum absorption (A) of 1.2. The cell length l is 1 cm, the concentration is 1.9 mg per 25 ml of solution and the molecular mass of the compound is 100 g/mol.

2) Calculate the molar absorption coefficient of a solution of concentration 10^{-4} M, placed in a 2 cm vessel, with $I° = 85.4$ and $I = 20.3$.

Exercise 3:

An aqueous solution of potassium permanganate ($C = 1.28.10^{-4}$ M) has a transmittance of 0.5 at 525 nm, using a cell with a 10 mm optical path.

1) Calculate the molar absorption coefficient of permanganate at this wavelength.

2) If the concentration is doubled, calculate the absorbance and transmittance of the new solution.

Exercise 4:

A 4 mm cell has been filled with a benzene solution. The benzene concentration is 1.0×10^{-3} mole.L^{-1} . The UV-visible spectrum of this solution shows a band at wavelength 256 nm.

1) Knowing that the transmittance of the sample is 59%, calculate the molar extinction coefficient of benzene at 256 nm.

2) What will be the absorbance at 256 nm of the same sample placed in a 2 mm cell?

Exercise 5:

Determination of chromium VI in water polluted to 0.1 ppm ($M = 52$ g/mol) is carried out indirectly by reaction with diphenylcarbazide as complexing agent to form a complex in an acid medium which absorbs at wavelength

λmax = 540 nm and has a molar extinction coefficient ε_{max} = 41700 L/mol.cm.

What value of optical path l must be used for the absorbance of the solution to be equal to 0.4?

Exercise 6:

Consider an aqueous solution S containing cobalt nitrate $Co(NO_3)_2$ and chromium nitrate $Cr(NO_3)_3$. The absorbances of three solutions are measured at two wavelengths, 510 nm and 575 nm, and the results obtained are summarized in the table below. The optical path of the cell used is of the order of 10 mm.

1. Calculate the molar extinction coefficients of each compound taken alone in aqueous solution at each wavelength

2. Calculate the molar concentrations C_X and C'_X of $Co(NO_3)_2$ and $Cr(NO_3)_3$ in solution S

Solution	λ_{max} (nm)	Absorbance A	Concentration
$Co(NO_3)_2$ alone in solution	510	0,7140	15.10^{-2} mol/L
	575	0,0097	
$Cr(NO_3)_3$ alone in solution	510	0,2980	6.10^{-2} mol/L
	575	0,7570	
Mix	510	0,4000	C_X and C'_X
	575	0,5770	

Exercise 7:

What are all the possible electronic transitions for the following molecules: CH_4 , $CH_3 Cl$, $H C=O_2$

Exercise 8:

The UV spectrum of acetone shows two absorption bands: λ_{max} = 280 nm with ε_{max} = 15 and λ_{max} = 190 nm with ε_{max} = 100. Identify the electronic transition in each of the two bands. Which is the most intense?

Exercise 9:

1) From the values of λ_{max} (in nm) for these molecules, what conclusions can be drawn about the relationship between λ_{max} and the structure of the absorbing molecule? Ethylene (170); Buta-1,3-diene (217); 2,3-Dimethybuta-1,3-diene (226); Cyclohexa-1,3-diene (256) and Hexa-1,3,5-triene (274).

2) Explain the following variations in the λmax (in nm) of the following compounds: CH_3 -X, when X=Cl (λ_{max} = 173), X=Br (λ_{max} = 204) and X=I (λ_{max} = 258).

Exercise 10:

1. Using the Woodward-Fieser rules, predict the maximum wavelengths λ_{max} of each of the compounds below taken in ethanol as solvent:

2. Using Scott's rules, predict the maximum wavelengths λ_{max} of each of the compounds below taken in ethanol as solvent:

3. Which compound has a bathchromic effect and which has a hypsochromic effect?

Exercise 11:

1) Which of the following aromatic compounds will absorb at the longest wavelength?

2) The UV spectrum ofα -cyperone, a naturally occurring ketone, shows a maximum at 254nm ($\varepsilon = 19000$). Two formulas have been proposed:

Which structure matches the UV spectrum?

Exercise 12:

Or the compound:

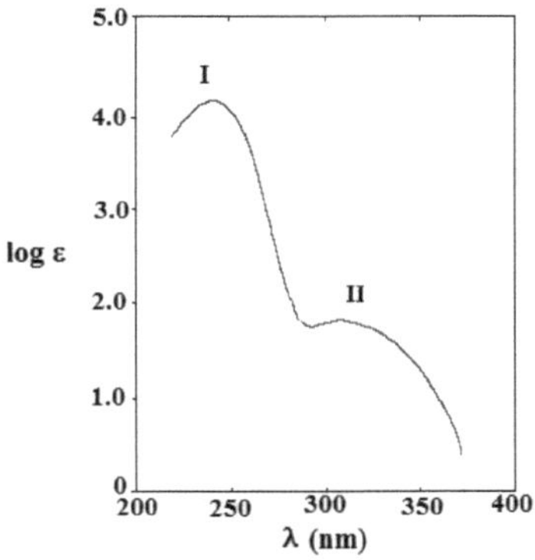

Its absorption spectrum as a function of wavelength in ethanol is shown below:

1- Give the values of the wavelengths and molar extinction coefficients of absorption bands I and II.

2- Assign each band to the corresponding transition, justifying your answer. Then specify the corresponding chromophore in each case.

3- What effect would using hexane as a solvent instead of ethanol have on the position of each band? Justify your answer using the corresponding energy level diagram.

Exercise 13:

1- In the case of the molecule[1] H -[35] Cl, we observe a band at 2990 cm^{-1} due to the fundamental vibration ($v = 0 \rightarrow v = 1$).

    1.a- Calculate the value of the force constant in the case of a harmonic oscillator.

1.b- Knowing that the isotopic effect has no influence on the strength constants of the Cl-H and Cl-D bonds. What would be the wave number absorbed by D-^{235}Cl?

2- The wave numbers of the bonds C=^{1216}O and C-^{1216}O are 1715 cm^{-1} and 1050 cm^{-1} respectively. Compare the stiffness of the bonds C=^{1216}O and C-O.^{1216}

Exercise 14:

Using Hooke's law (for the harmonic vibrator), give the increasing order of the vibrational wavenumbers of the C-X bond for X= Br, F, Cl. Assume that the force constants have the same value.

Exercise 15:

IR spectroscopy is ideal for confirming the presence of functional groups. Let the reaction be :

1) BH$_3$

2) H$_2$O$_2$/ NaOH

hexène — hexanol (OH)

Describe the main differences expected in the IR spectra of the reactant (hexene) and the product (hexanol): what characteristic vibrations appear and disappear?

Exercise 16:

1- The IR spectrum of acetone A shows strong absorption at 1715 cm .$^{-1}$

O

Which acetone vibration corresponds to this transition?

2- The wave number associated with the same vibration in molecule B is equal to 1670 cm .$^{-1}$

O

What explains this move?

Exercise 17:

Consider the infrared spectra 1-3 shown below. Each corresponds to a compound from the following list: pentan-2-one, hex-1-yne, phenylmethanol.

Assign the corresponding compound to each spectrum, indexing the most important bands.

Spectrum 1

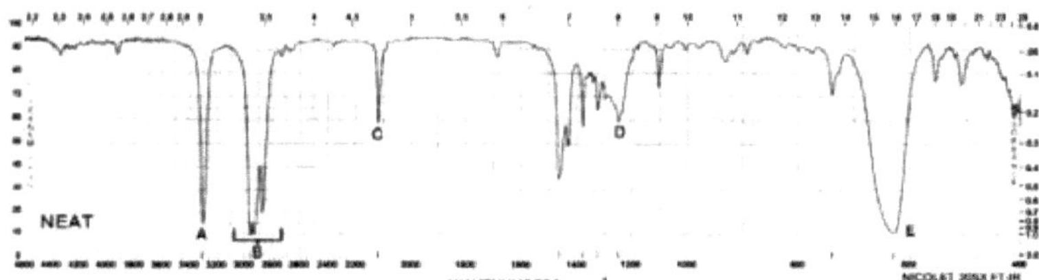

Spectrum 2

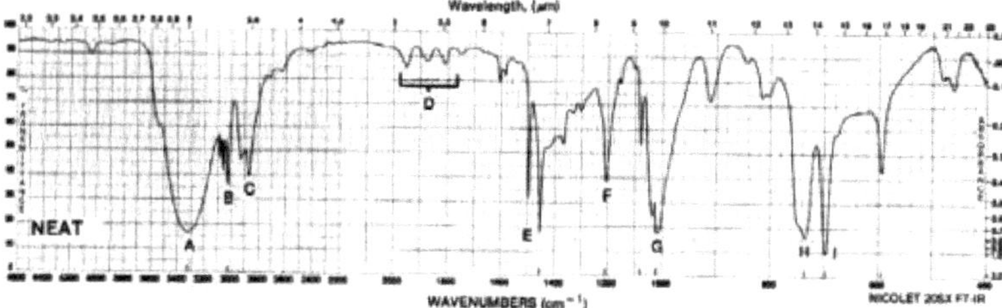

Spectrum 3

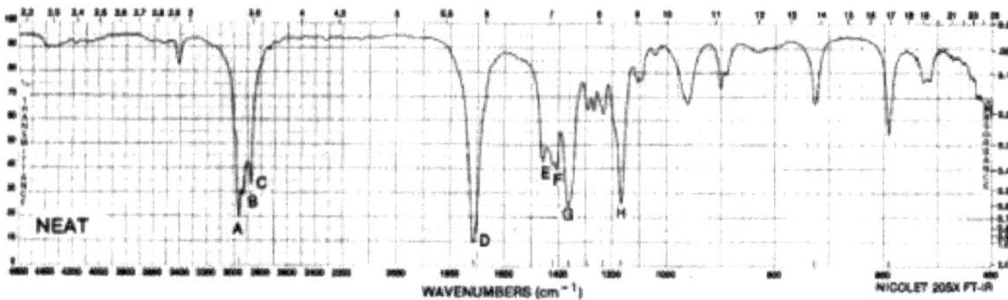

Exercise 18:

Explain why hydrogen bromide is IR-active while bromine is IR-inactive.

Exercise 19:

Figures A and B show extracts from the infrared spectra of compounds 1 and 2 respectively. The spectrum of compound 1 was obtained from a film of pure 1 in the liquid state, while that of compound 2 was obtained from a dilute solution of 2 in tetrachloromethane (CCl_4).

Interpret and assign the following IR spectra to the compounds below. Justify your answer by indicating on the spectra the assignment of the IR absorption bands, noted a, c, d, g, i and j, characteristic of the bonds present in the molecules of 1 and 2.

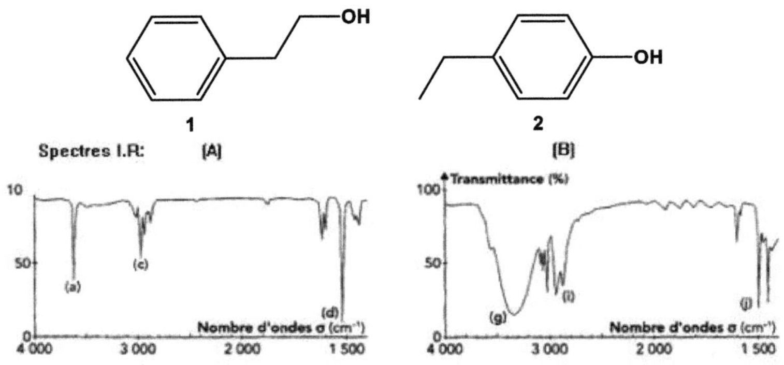

Exercise 20:

Assign the corresponding compound to each spectrum (1 to 3) (justify your answer by analyzing the characteristic bounds).

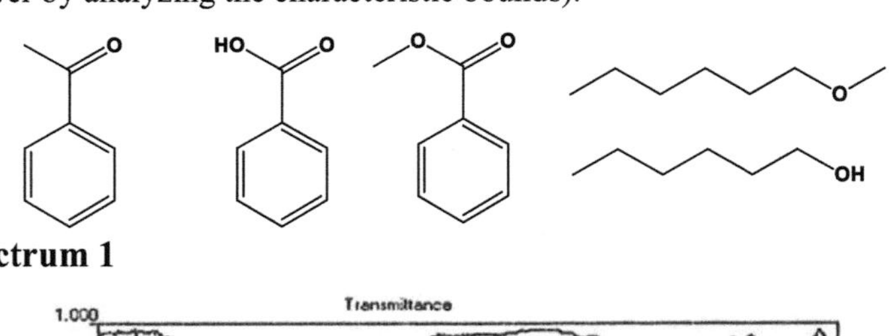

Spectrum 1

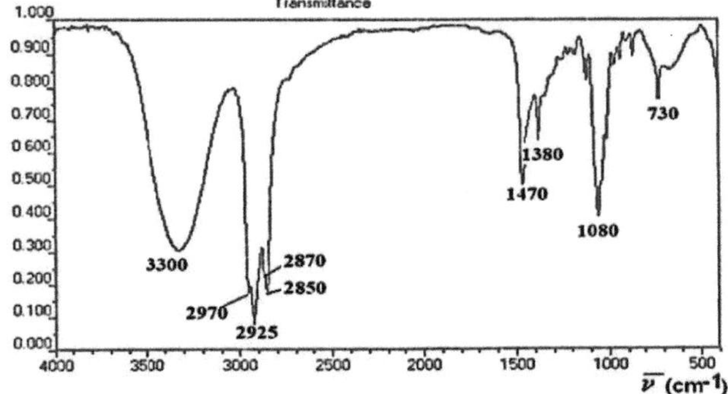

Spectrum 2

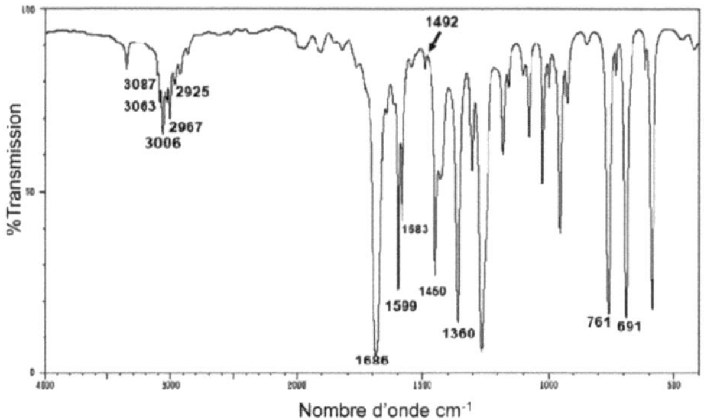

Spectrum 3

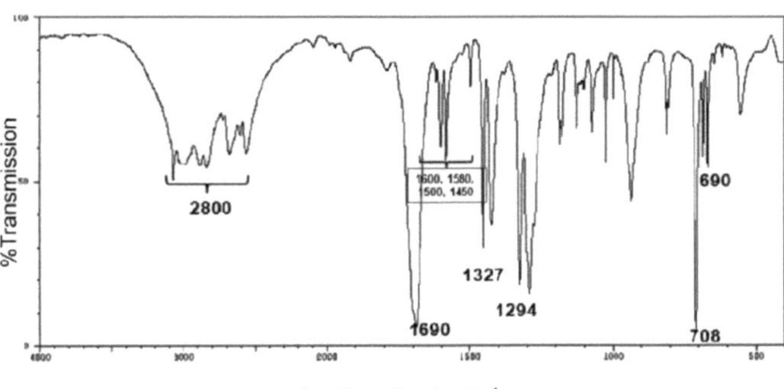

Nombre d'onde cm⁻¹

Exercise 21:

Assign the corresponding compound to each spectrum (justify your answer by analyzing the characteristic bounds).

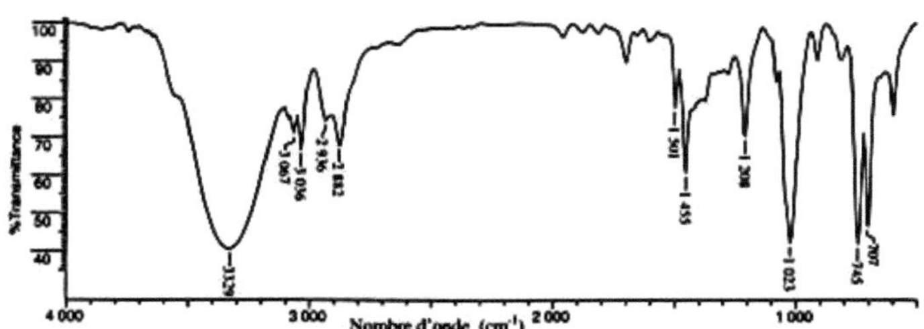

Spectrum 1

Spectrum 2

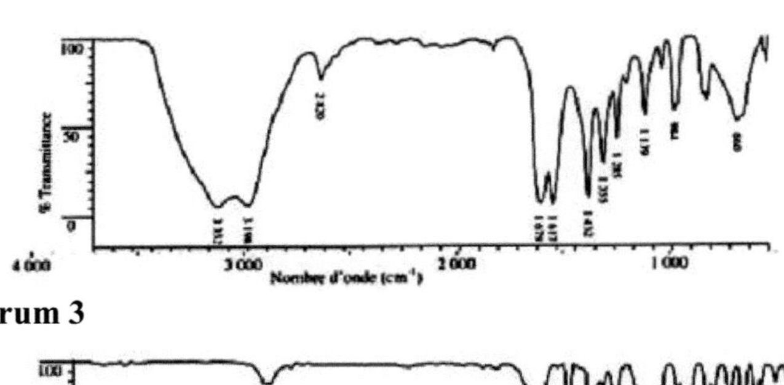

Spectrum 3

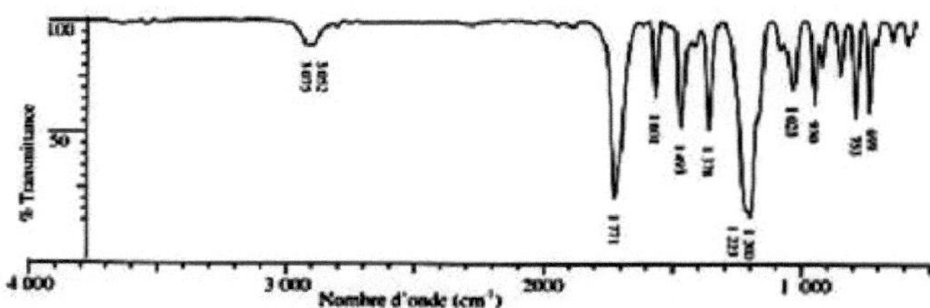

Spectrum 4

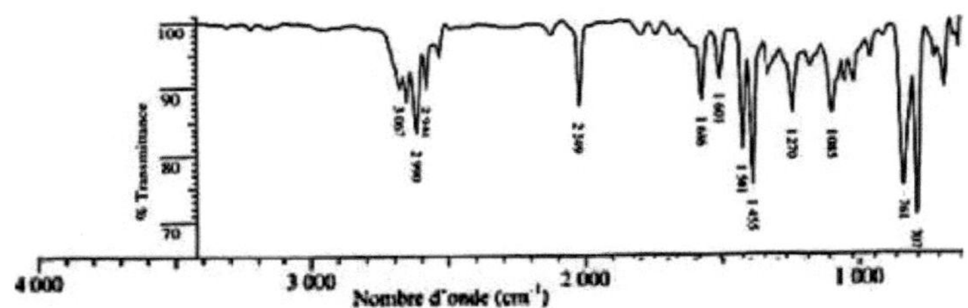

Solving the exercises

Exercise 1:

Calculate the mass of aniline needed to prepare 100 ml of such a solution?

λ_{max} = 280 nm; ε_{max} =1430 L. mol-1. cm $^{-1}$

% transmission = 30 → T = 0.3

l = 1 cm and V= 100 mL

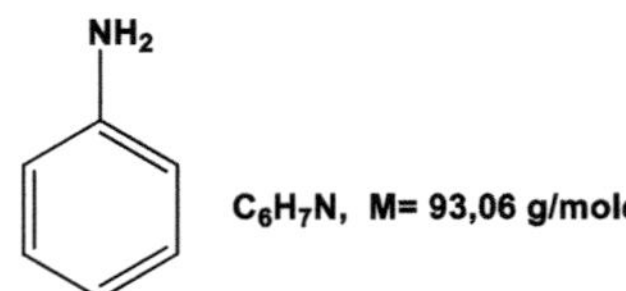

$M = 6 \times 12 + 7 \times 1 + 14 = 93 \ g/mol$

$A = -\log_{10} T = \square. \square. c \rightarrow c = -\log_{10} T \ (\square. \square)^{-1}$

$c = n \ v^{-1} = m \ (M \times \square)^{-1} = -\log_{10} T \ (\square. \square)^{-1} \rightarrow m = -M \times v \times \log_{10} T \ (\square. \square)^{-1}$

$m = -93 \times 100.10^{-3} \times \log_{10} 0.3 \ (1430.1)^{-1} = 3.4.10\text{-}3g$

$$m = 3.4 \ mg$$

Exercise 2:

1) Beer Lambert's law is applied, **ε= 1578.94 mol-1 .l.cm .$^{-1}$**

2) ε = **3119.8 mol^{-1} .l.cm .$^{-1}$**

Exercise 3:

1) Beer Lambert's law is applied, ε = **2351.7 mol^{-1} .l.cm .$^{-1}$**

2) **A = 0.601 and T = 0.25.**

Exercise 4:

A 4 mm cell has been filled with a benzene solution. The benzene concentration is 1.0×10^{-3} mole.L^{-1} . The UV-visible spectrum of this solution shows a band at wavelength 256 nm.

1) Knowing that the transmittance of the sample is 59%, calculate the molar extinction coefficient of benzene at 256 nm.

Beer-Lambert's law applies: $A = \log (I_0 /I) = -\log T = \varepsilon \ l \ C$

A: Absorbance

I_0 : incident light intensity

I: intensity of transmitted light

T: transmission

ε: molar extinction coefficient, L.mol.$^{-1}$ .cm^{-1}

l: optical path, cm

C: molar concentration, mol.L^{-1}

Transmittance is given. Transmittance is %T. We are looking for ε:

$$\varepsilon = \frac{-\log T}{l\,C}$$

A.N: $\varepsilon = -\log 0.59/0.4\,10^{-3}$

$$\varepsilon = 572.87 \text{ L.mol.}^{-1} \text{ .cm}^{-1}$$

2) What will be the absorbance at 256 nm of the same sample placed in a 2 mm cell?

A= ε l C

We know that at a given wavelength, ε is a constant.

A.N. at $\lambda = 256$ nm, $\varepsilon = 572.87$ L.mol.$^{-1}$ .cm^{-1}

A = 572.87 x 0.2 x 10^{-3}

$$A = 0{,}115$$

For the same solution (constant C), doubling the value of l doubles the value of A.

Exercise 5:

Data: C = 0.1.10-3 g.L-1 , λmax = 540 nm , A = 0.40

Molar absorption coefficient: $\varepsilon = 41700$ L.mol^{-1} .cm^{-1}

Before applying the Beer-Lambert law, the units of the parameters used in the formula must be checked:

The concentration of chromium in the polluted water must be expressed in mol.L^{-1} (molar concentration), so divide by the molar mass :

C = 0.1.10^{-3} /52 = 1.92.10^{-6} mol.L^{-1}

According to Beer-Lambert's law, for a maximum wavelength one: A = $\varepsilon.l.$c

Hence: l = A/ ε..c

l represents the optical path of the cuvette

Numerical application:

$l = 0.4 / (41700.1,92.10^{-6}) = 4.99$ cm about 5 cm

Exercise 6:

Calculate molar absorption coefficients $\varepsilon Co(510)$, $\varepsilon Cr(510)$, $\varepsilon Co(575)$ and $\varepsilon Cr(575)$:

a- Beer Lambert's law is applied:

$$\varepsilon_{Co(510)} = 4.\ 76 \text{ and } \varepsilon_{Co\ (575)} = 0.64$$

$$\varepsilon_{Cr\ (510)} = 4.\ 96 \text{ and } \varepsilon_{Cr\ (575)} = 12.61$$

b- Molar concentrations (mol-L^{-1}) of the two salts **A** and **B** in the sample solution.

The additivity of absorbances is applied:

At 510 nm, we have $0.4 = \varepsilon Co(510)*Cx + \varepsilon Cr(510)*Cx'$.

At 575 nm, we have $0.5770 = \varepsilon Co(575)*Cx + \varepsilon Cr(575)*Cx'$.

$Cx = 0.0438$ mol/L ; $Cx' = 0.0384$ mol/L

$$C_A = 1.2.10^{-1} \text{ mol.L}^{-1} \text{ and } C_B = 2.10^{-2} \text{ mol.L}^{-1}$$

Exercise 7:

The transitions are :

CH$_4$: transition $\sigma \rightarrow \sigma^*$

CH$_3$ Cl : $\sigma \rightarrow \sigma^*$ and $\quad$ n $\rightarrow \sigma^*$

CH$_2$ O : $\sigma \rightarrow \sigma^*$ transitions, $\quad$ n $\rightarrow \sigma^*$, n $\rightarrow \pi^*$ and $\pi \rightarrow \pi^*$ transitions

Exercise 8:

1) $\lambda = 280$ nm: **n $\rightarrow \pi^*$ transition**

$\lambda = 190$ nm: **$\pi \rightarrow \pi^*$ transition**

2) The most intense transition is: $\quad$ **$\pi \rightarrow \pi^*$**

Exercise 9:

1) We can conclude that λ increases with increasing carbon chain and conjugation.

A cyclic compound absorbs at a higher λ than its aliphatic counterpart.

2) This is the **n $\rightarrow \pi^*$ transition.**

As electronegativity decreases, the transition becomes easier and λ increases.

Exercise 10:

1)

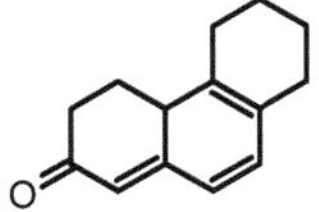

		Increment added (nm)
Basic structure	Hetero-annular diene	214
Substituent	2 exo-cyclic double bond	10
	3 alkyl residue	10
	1 alkoxy	6
λ_{max} = 240 nm		

		Increment added (nm)
Basic structure	Diènehomo-annulaire	253
Substituent	1 exo-cyclic double bond	5
	3 alkyl residue	10
	Additional conjugated double bond	30
λ_{max} = 298 nm		

	Increment added (nm)
Carbonyl compound □, □□insaturé	215
1 exo-cyclic double bond	5
Alkyl in β	12
Alkyle at δ +1	18
2 Alkyls at δ +2	36
Homo-annular diene	39
2 Additional conjugated double bonds	60
λ_{max} = 385 nm	

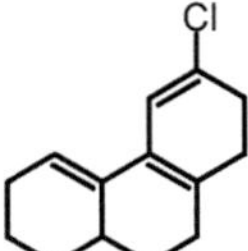

	Increment added (nm)
Diènehomo-annulaire	253
1 exo-cyclic double bond	5
4 remaining Alkyle	20
Cl	5
λ_{max} = 283 nm	

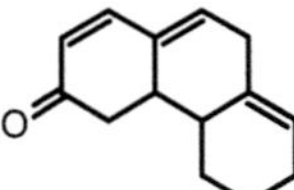

	Increment added (nm)
Carbonyl compound α, β insaturé	215
2 exo-cyclic double bond	10
Alkyle en γ	18
Alkyle en δ	18
Additional conjugated double bond	30
λ_{max} = 291 nm	

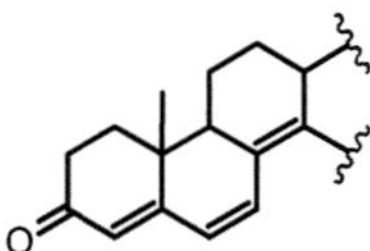

	Increment added (nm)
Carbonyl compound α, β insaturé	215
2 exo-cyclic double bond	10
Alkyl in β	12
Alkyle at δ +1	18
2 Alkyls at δ +2	36
2 Additional conjugated double bond	60
λ_{max} = 353 nm	

	Increment added (nm)
Carbonyl compound α, β insaturé	193
Alkyle en α	10

Alcoxy in β	30
λ_{max} = 233 nm	

2/

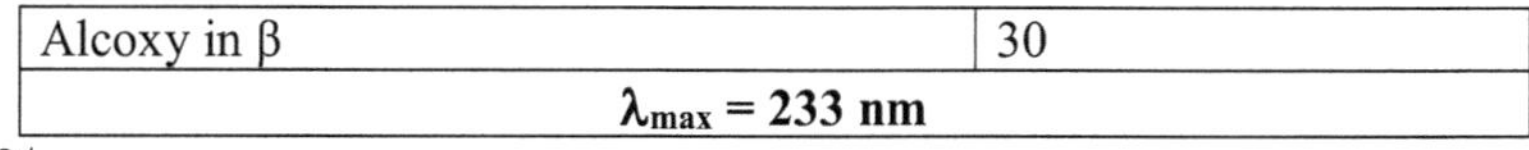

	Increment added (nm)		
Basic structure	230	246	250
Alkyle in ortho	3	3	
OH in meta	-	7	7
Cl in para	-	-	10
NH2 in meta	15	-	-
λ_{max}	**248 nm**	**256 nm**	**267 nm**

3/ Bathochromic aldehyde, hypsochromic ester

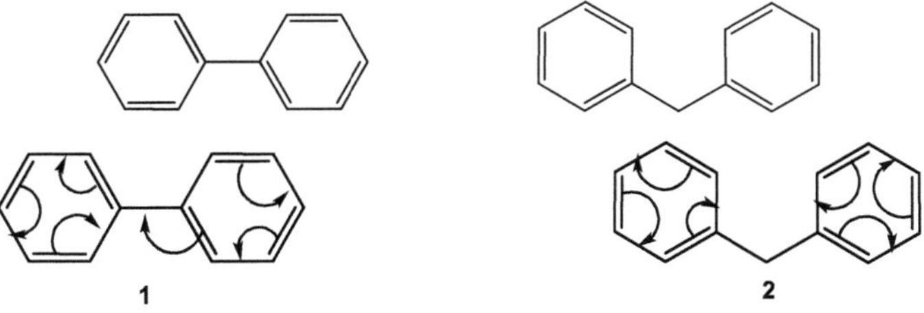

Exercise 11:

1- Which of the following aromatic compounds will absorb at the longest wavelength?

Conjugation between the two   No   conjugation between the two

aromatic cycles   aromatic cycles

The conjugation of compound **1** is more extensive than that of compound **2**. Conjugation causes a bathochromic effect on the $\pi \rightarrow \pi^*$ transition.

$\lambda_1 > \lambda_2$

Compound **1** will absorb at the longer wavelength thanks to the conjugation of the two phenyls.

2- The UV spectrum of α -cyperone, a naturally occurring ketone, shows a maximum at **254** nm (ε = 19000). Two formulas have been proposed:

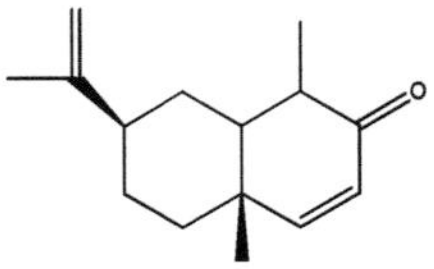

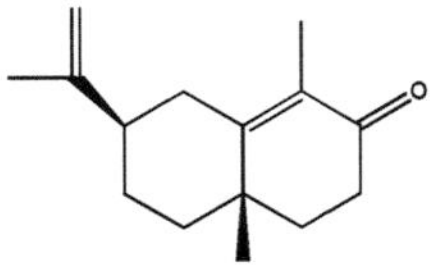

Which structure matches the UV spectrum?

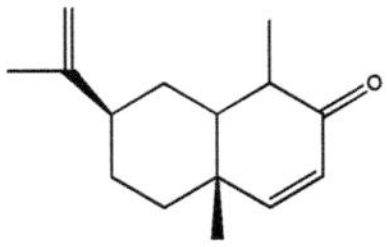

	Increment added (nm)
Carbonyl compound □, □insaturé	215
exo-cyclic double bond	5
Alkyl in β	12
λ_{max} = 232 nm	

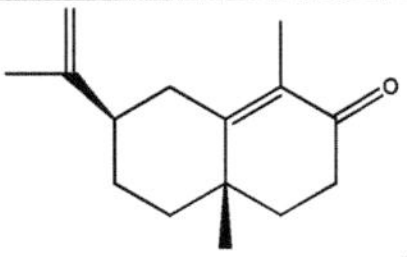

	Increment added (nm)
Carbonyl compound □, □□insaturé	215
exo-cyclic double bond	5
Alkyl in α	10
2 β-Alkyl	12x2
λ_{max} = 254 nm	

The structure:

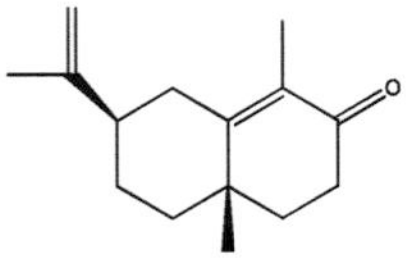

is consistent with the UV spectrum

Exercise 12:

Wavelength values and molar extinction coefficients for absorption bands I and II.

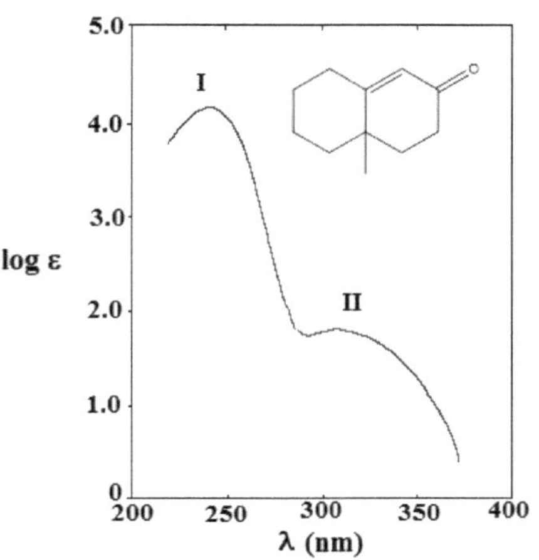

- Band I: λ_{max} = 245 nm, $\log\varepsilon$ = 4.12, ε_{max} = 13300 L.mol^{-1}.cm^{-1}

- Band II: λ_{max} = 310 nm, $\log\varepsilon$ = 1.8, ε_{max} = 63.1 L.mol^{-1} .cm^{-1}

2- Allocation of bands I and II to each transition.

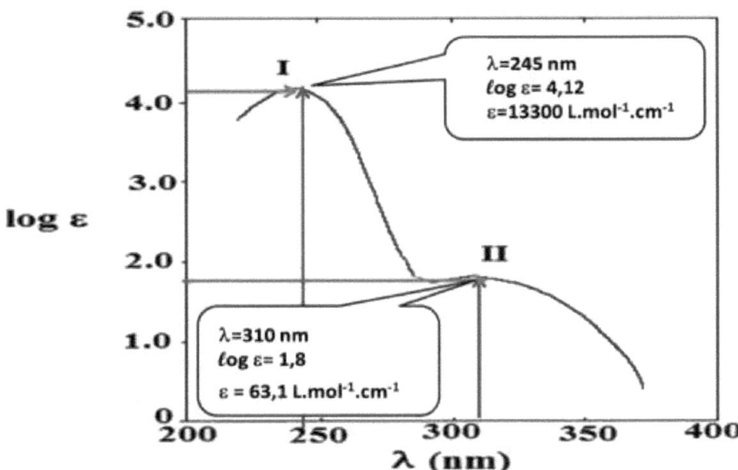

- Band I: ε_{max} = 13300 L.mol^{-1} .cm^{-1} > 1000

Transition $\pi \rightarrow \pi^*$ ⟶ Chromophore C=C

- Band II: ε_{max} = 63.1 L.mol^{-1} .cm^{-1} < 100

Transition n $\rightarrow \pi$ * $\longrightarrow$ Chromophore C=O

3- The effect of using hexane as a solvent instead of ethanol on the position of each band.

• Transition $\pi \rightarrow \pi$ *

Moving from ethanol (polar solvent) to hexane (apolar solvent) (polarity decreases), the energy of the excited state *increasesπ

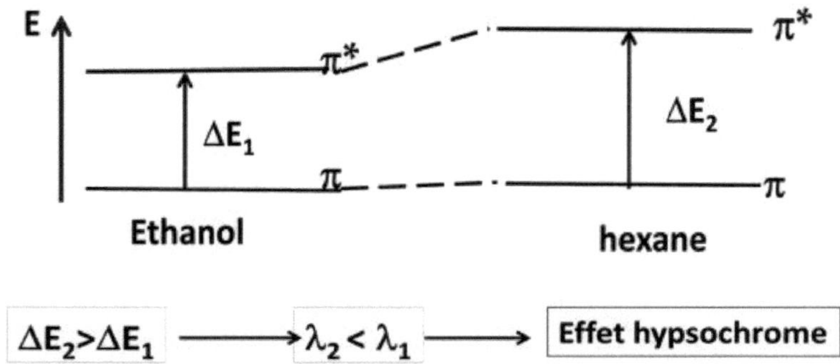

• Case of transition n $\rightarrow \pi$ *

The state is stabilized under the effect of the polar solvent by the formation of hydrogen bonds.

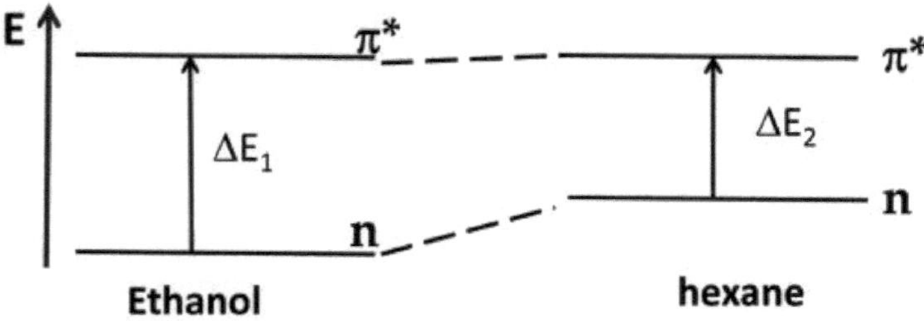

Moving from ethanol (polar solvent) to hexane (apolar solvent) (polarity decreases), the energy of the fundamental n state increases.

$$\Delta E_2 < \Delta E_1 \quad \longrightarrow \quad \lambda_2 > \lambda_1 \quad \longrightarrow \quad \boxed{\textbf{Effet Bathochrome}}$$

Exercise 13:

In the case of the molecule[1] H -[35] Cl, a band is observed at 2990 cm[-1] due to the fundamental vibration (v = 0 → v = 1).

1.a- Calculate the value of the force constant for a harmonic oscillator.

The molecule[1] H -[35] Cl is considered as a harmonic oscillator. We can then write:

$$\bar{v}_{HCl} = \frac{1}{2\pi c} \sqrt{\frac{k_{HCl}}{\mu_{HCl}}}$$

with

$$\mu_{HCl} = \frac{m_H \times m_{Cl}}{m_{Cl} + m_H}$$

In the C.G.S. system :

$\bar{v}$ wave number in cm[-1]

c: celerity in cm.s [-1]

k: link force constant in dyn.cm[-1]

μ: reduced mass in grams

m_H and m_{Cl} : masses in grams of the H and Cl atoms respectively

Hence the force constant

$$\boxed{k_{HCl} = 4\,\pi^2 c^2 \bar{v}^2\, \mu_{HCl}}$$

Reduced weight calculation

$$\frac{1}{\mu_{HCl}} = \frac{1}{m_H} + \frac{1}{m_{Cl}} \rightarrow \mu_{HCl} = \frac{m_H \times m_{Cl}}{m_{Cl} + m_H} \quad avec\ m_H = \frac{M_H}{\mathcal{N}}\ et\ m_{Cl} = \frac{M_{Cl}}{\mathcal{N}}$$

$$\rightarrow \mu_{HCl} = \frac{M_H \times M_{Cl}}{\mathcal{N}(M_{Cl} + M_H)} \qquad \rightarrow \mu_{HCl} = \frac{1 \times 35}{6,02.10^{23}\,(1+35)} = \frac{35}{216,72.10^{23}}$$

$$\boxed{\mu_{HCl} = 0,161498.10^{-23}\ g}$$

Calculation of the force constant k $_{HCl}$

$$k_{HCl} = 4\,\pi^2 c^2 \bar{v}^2\, \mu_{HCl}$$

$$\rightarrow k_{HCl} = 4 \times (3.14)^2 \times (3.10^{10})^2 \times (2990)^2 \times 0{,}161498.10^{-23}$$

$$\boxed{k_{HCl} = 0{,}512.10^6\ g.s^{-2}}$$

In the C.G.S. system the dyne :

$$\mathbf{g.\,s^{-2}.\,cm} \quad \rightarrow \quad k_{HCl} = 5{,}12.10^5\ dyne.\,cm^{-1}$$

In the SI international system, force is expressed in N where

$$1\ N = Kg.\ m.\ s^{-2} = 10^3\ g.\ 10^2$$

$$cm.\ s^{-2} = 10^5\ g.cm.\ s^{-2} = 10^5\ dyne$$

Hence

$$k_{HCl} = 5{,}12.\,10^5\ dyne.\,cm^{-1} \rightarrow \boxed{k_{HCl} = 5{,}12\ N.\,cm^{-1} \rightarrow k_{HCl} = 5{,}12.10^2\ N.\,m^{-1}}$$

1.b- Calculation of the wave number absorbed by^2 D -^{35} Cl .

$$\bar{v}_{DCl} = \frac{1}{2\pi c}\sqrt{\frac{k_{DCl}}{\mu_{DCl}}} \qquad \text{avec} \qquad \mu_{DCl} = \frac{m_D \times m_{Cl}}{m_{Cl} + m_D}$$

et on suppose que $k_{DCl} = k_{HCl} = 5{,}12.\,10^5\, dyne.\,cm^{-1}$

$$\mu_{DCl} = \frac{70}{6{,}02 \cdot 10^{23}\ \times 37} = 0{,}314.\,10^{-23}\ g$$

$$\mu_{DCl} = 0{,}314.\,10^{-23}\ g$$

$$\bar{v}_{DCl} = \frac{1}{2\times3{,}14\times3.10^{10}}\sqrt{\frac{5{,}12.10^5}{0{,}314.10^{-23}}}$$

$$\bar{v}_{DCl} = 0{,}2139.\,10^4 \cong 0{,}214.\,10^4\ cm^{-1}$$

$$\rightarrow \boxed{\bar{v}_{DCl} = 2140\ cm^{-1} \qquad \bar{v}_{HCl} = \sqrt{2}\ \bar{v}_{DCl}}$$

When H is replaced by D, $\bar{v}$ is shifted to lower wavenumbers as μ increases.

2- Compare the stiffness of C=^{1216}O and C-^{1216}O bonds.

For C=O

$$v(c = o) = \frac{1}{2\pi}\sqrt{k(c = o)/\mu}$$

$$\nu^2(c=o) = \frac{1}{4\pi^2}k(c=o)/\mu$$

For C O-

$$\nu(c-o) = \frac{1}{2\pi}\sqrt{k(c-o)/\mu}$$

$$\nu^2(c-o) = \frac{1}{4\pi^2}k(c-o)/\mu$$

$$\frac{\nu^2(c=o)}{\nu^2(c-o)} = \frac{k(c=o)}{k(c-o)}$$

Gold $\nu(c=o) > \nu(c-o)$

So $k(c=o) > k(c-o)$

$$\mu(c=o) = \frac{12x16}{12+16} = 6{,}86\,\frac{g}{mol},$$

$$\mu(c=o) = \frac{6{,}86}{mol}x\frac{1mol}{6{,}022\,.10^{23}}x\frac{1kg}{1000g} = 11{,}4\,.\,10^{-27}kg = \mu(c-o)$$

$$\nu^2(c=o) = \frac{1}{4\pi^2}k(c=o)/\mu \Leftrightarrow k(c=o) = 4\pi^2\mu\nu^2(c=o)$$

$$v = c/\lambda \qquad \text{and} \quad v = c.\,\sigma$$

$$\nu(\text{c=o}) = 3.10^8 \text{ m/s} \,/(\frac{1cm}{1715}x\frac{1m}{100cm}) = 5.145\ 10^{13}\ \text{Hz}$$

$$k(c=o) = 1191{,}3\,kg.\,Hz^2$$

$$k(c=o) = 1191{,}3\,kg/S^2$$

$$k(c=o) = 1191{,}3\,N/m$$

$$\nu(\text{c-o}) = 3.10^8 \text{ m/s} \,/ (\frac{1cm}{1050}x\frac{1m}{100cm}) = 3.15\ 10^{13}\ \text{Hz}$$

$$k(c-o) = 446{,}6\,N/m$$

So $k(c=o) > k(c-o)$

Exercise 14:

Using Hooke's law (for the harmonic vibrator), give the increasing order of the vibrational wavenumbers of the C-X bond for X= Br, F, Cl. Assume that the force constants have the same value.

The C-X bond is considered as a harmonic oscillator. We can then write :

The practical quantity in vibrational spectroscopy is the wave number.

$$\overline{v}\,[cm^{-1}] = \frac{1}{2\pi c}\sqrt{\frac{k}{\mu}}$$

In the C.G.S. system : $\overline{v}$: wave number in cm^{-1} ; c : celerity in cm.s^{-1}

k: bond force constant in dyn.cm^{-1} ; μ : reduced mass in grams

$\overline{v}$ depends: - the reduced mass μ of the A-B system (vibration frequency inversely proportional to μ)

- the force constant of the bond

Link	m_2 (g)	$\overline{v}$ (cm)$^{-1}$	μ (g)	
C-H	1	3300	0,923	
C-C	12	1200	6,000	
C-O	16	1100	6,857	
C-Cl	35,5	800	8,968	
C-Br	80	550	10,43	
C-I	127	500	10,96	

As the reduced mass μ increases, the vibrational wavenumber of the C-X bond for X= Br, Cl and F decreases.

Vibration frequency inversely proportional to μ:

Exercise 15:

IR spectroscopy is ideal for confirming the presence of functional groups. Let the reaction be :

1) BH$_3$

2) H$_2$O$_2$/ NaOH

hexène → hexanol   OH

▶ Vibrations characteristic of the Hexene molecule:

Groupements	vibrations	Nombre d'onde (cm^{-1})
C=C (alcène)	$\nu_{C=C}$	1645
CH=CH$_2$ (vinyle)	$\nu_{=CH}$	3095-3075 3040-3010
	$\gamma_{=CH}$	995-985 915-905

► Vibrations characteristic of the 1-Hexanol molecule:

Groupements	vibrations	Nombre d'onde (cm^{-1})
OH (alcool)	v_{O-Hass}	(3400-3200) Bande large
C-O (alcool primaire)	v_{C-O}	(1085-1050) Forte

IR spectroscopy concludes that alcohol formation has taken place if the bands due to $v_{=CH}$, $v_{C=C}$ and $\gamma_{=CH}$ of the vinyl group disappear and the bands due to v_{OH} and v_{C-O} of the alcohol's OH group appear.

Exercise 16:

1- The IR spectrum of acetone A shows strong absorption at 1715 cm $.^{-1}$

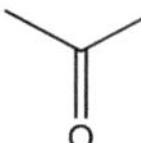

The strong absorption at 1715 cm^{-1} corresponds to the valence vibration of the C=O carbonyl group ($v_{C=O}$) of the acetone molecule.

2- The wave number associated with the same vibration in molecule B is equal to 1670 cm $.^{-1}$

The decrease in the wave number of the valence vibration $v_{C=O}$ of molecule B is due to the conjugation of C=O with C=C.

Exercise 17:

Using the tables, we can easily identify compounds based mainly on functional groups, we should find :

<u>Spectrum 1: hex1-yne ($\bar{v}$ ($\equiv C_{sp}$ -H) = 3310 cm^{-1} , $\bar{v}$ ($C_{sp} \equiv C_{sp}$) = 2119 cm</u> <u>)$.^{-1}$</u>

Note the weak band of elongation $C_{sp} \equiv C_{sp}$ at 2110 cm^{-1} . It is not always visible, especially in the case of disubstituted alkynes.

In contrast, the $\equiv C_{sp}$ -H elongation band of monosubstituted alkynes is always intense, emerging here at 3268 cm $.^{-1}$

Less important to report is the deformation band of $\equiv C_{sp}$ -H acetylenic (630 cm^{-1}), as well as its first harmonic (1247 cm)$.^{-1}$

Spectrum 2: phenylmethanol ($\bar{\upsilon}$ (OH) = 3330 cm)$^{-1}$

For a molecule diluted in an apolar aprotic solvent, i.e. when there are no H bonds, the frequency is between 3600 and 3584 cm .$^{-1}$

On the other hand, for pure benzyl alcohol, with many strong H bonds, this frequency drops to 3300 cm .$^{-1}$

$_{O-H}$ ☐to 1208 cm^{-1} and υ_{C-O} to 1017 cm^{-1}

Let's leave aside the strips already studied (υ_{C-H} , δ_{C-H} , γ_{C-H} ☐.

Hypsochrome for υ_{CAr-H} (3045 cm^{-1}), δ_{O-H} (1360 cm^{-1}) and υ_{CAr-O} (1223 cm^{-1}), and batochrome for $\upsilon_{CAr-CAr}$ (1580 cm^{-1}). The two bands γ_{CAr-H} ☐pour monosubstitution are found at 685 and 745 cm^{-1} (G and H).

Spectrum 3: pentan-2-one ($\bar{\upsilon}$ (C=O) = 1717 cm)$^{-1}$

All organic compounds containing a C=O carbonyl group have a characteristically intense absorption at around 1700 cm^{-1} : this is the most intense and sharpest band in an IR spectrum.

The value of C=O absorption depends on the physical state (solid, liquid, vapor, solution), the effects of neighboring groups, conjugation and any H-bonds.

An aliphatic ketone absorbs around 1715 cm .$^{-1}$

On the two ketone spectra proposed, we will find them at 1725 cm^{-1} (unconjugated) and 1683 cm^{-1} (conjugated) respectively.

Note the existence of a weak C-CO-C elongation band, at 1172 cm^{-1} for the first compound, at 1255 cm^{-1} , stronger, for the aromatic ketone.

Exercise 18:

Explain why hydrogen bromide is IR-active while bromine is IR-inactive.

Not all vibration movements are active in infrared.

It's important to note, however, that a bond with zero dipole moment will give no signal in infrared spectroscopy.

Thus, Br-Br dibromium will not be active - its dipole moment being zero due to the symmetry of the molecule, it will have no active infrared bands.

On the other hand, Br-H hydrogen bromide will show a signal due to the Br-H bond, its dipole moment being non-zero.

Exercise 19:

Figures A and B show extracts from the infrared spectra of compounds **1** and **2** respectively. The spectrum of compound **1** was obtained from a film of pure 1 in the liquid state, while that of compound **2** was obtained from a dilute solution of **2** in tetrachloromethane (CCl)$_4$

Interpret and assign the following IR spectra to the compounds below. Justify your answer by indicating on the spectra the assignment of the IR absorption bands, noted a, c, d, g, i and j, characteristic of the bonds present in the molecules of **1** and **2**.

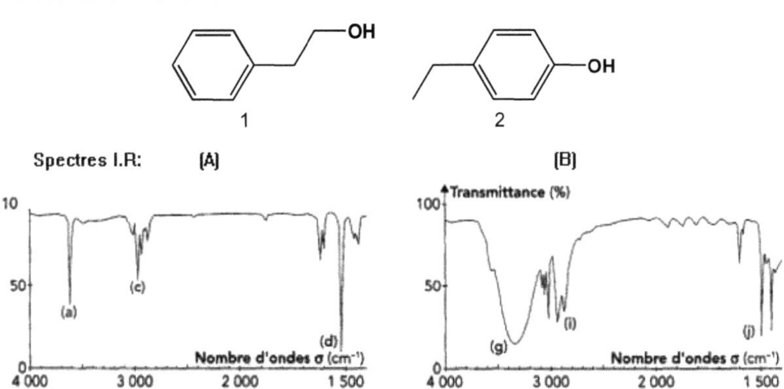

The spectrum of compound **2** was obtained from a dilute solution of **2** in tetrachloromethane (CCl$_4$): broad band disappears and a fine band appears, zone 3590-3650 cm^{-1} : free vOH which corresponds to spectrum (A).

The spectrum of compound 1 was obtained from a film of pure **1** in the liquid state: broad band between 3200 cm^{-1} and 3400 cm^{-1} and OH associated by hydrogen bonds vOH associated which corresponds to spectrum (B).

♣ Spectrum (A): corresponds to compound **2**;

(a): thin band 3600 cm^{-1} corresponds to free OH

(c): thin band 3000 cm^{-1} corresponds to aromatic CH

(d): thin band 1520 cm^{-1} corresponds to aromatic C=C

♣ Spectrum (B): corresponds to compound **1**;

(g): 3600 cm wide band^{-1} corresponds to OH linked by hydrogen bonds

(i): thin band 2950 cm^{-1} corresponds to aromatic CH

(j): thin band 1500 cm^{-1} corresponds to aromatic C=C

Exercise 20:

Assign the corresponding compound to each spectrum (1 to 3) (justify your answer by analyzing the characteristic bounds).

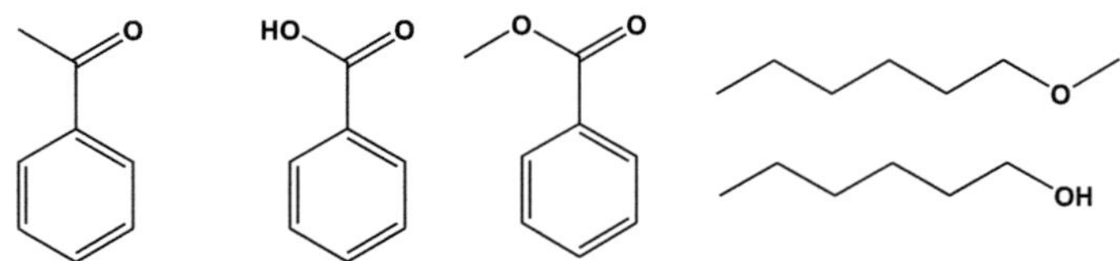

Spectrum 1

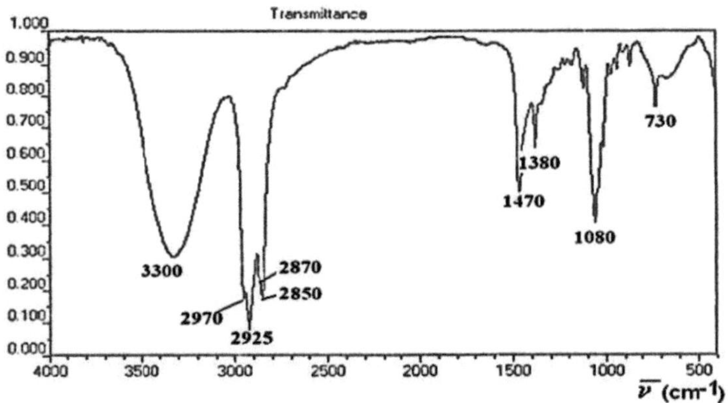

Region: 4000 -2000 cm^{-1}

Number of waves cm^{-1}	Intensity	Allocation
3300	F, wide	υ_{O-H} Ασσοχι
2970		υ^a_{CH3}
2925	TF	υ^a_{CH2}
2870		υ^a_{CH3}
2850	F	υ^a_{CH2}

Number of waves cm⁻¹	Intensity	Allocation
1470	F	δ^a_{CH3}, δ_{CH2}
1380	M	δ^s_{CH3}
1080	F	$_{C-O}$ ☐Primary alcohol
730	m	$_{CH2}$: ρ (XH$_2$ ☐n> 4

Formula:

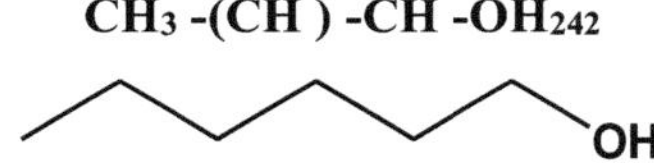

CH$_3$ -(CH$_2$)$_4$ -CH$_2$ -OH

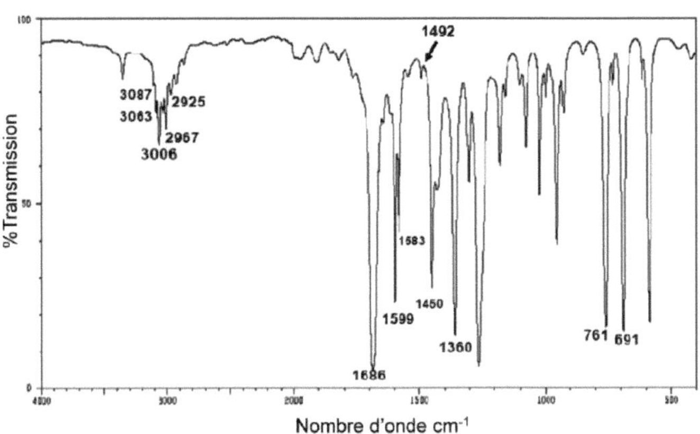

Spectrum 2

Region 4000-2000 cm⁻¹

Nombres d'onde cm⁻¹	Intensité	Attribution
3087, 3063, 3006	variables	$\nu_{=CH\ aromatique}$
2967	TF	ν^a_{CH3}
2925	TF	ν^s_{CH3}

Region 2000 - 400 cm⁻¹

Nombres d'onde cm^{-1}	Intensité	Attribution
1686	TF	$\nu_{C=O}$ conjuguée
1599, 1583, 1492, 1450	variables	$\nu_{C=C}$ aromatique
1430	F	δ^a CH3
1360	F	δ^s CH3
761	TF	$\gamma_{=CH}$ aromatique
691	TF	

The two strong bands at 761 and 691 cm^{-1} indicate the presence of a monosubstituted aromatic ring.

Formula:

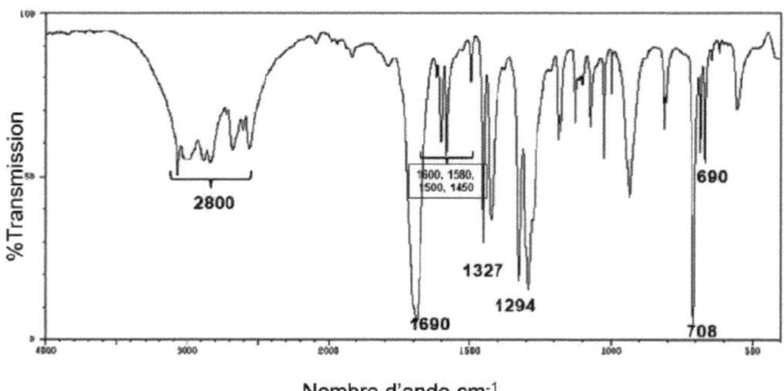

Acétophénone

Spectrum 3

Region 4000-2000 cm^{-1}

Strong, wide, structured band between 3400 and 2400 cm^{-1} , centered around 2800 cm $^{-1}$

For vibrations ν_{CH} aromatic between 3080 and 3030 cm^{-1} , they are masked by the strong, wide Band.

Region 2000 - 400 cm^{-1}

Nombres d'onde cm^{-1}	Intensité	Attribution
1690	TF	$\nu_{C=O\ conjuguée}$
1600, 1580, 1500, 1450	variables	$\nu_{C=C\ aromatique}$
1327	M	δ_{OH}
1294	F	ν_{C-O}
708	TF	
690	M	$\gamma_{=CH\ aromatique}$

Two strong bands at 708 cm^{-1} and 690 cm^{-1} 5H adjacent to a monosubstituted aromatic ring.

Exercise 21:

Assign the corresponding compound to each spectrum (justify your answer by analyzing the characteristic bounds).

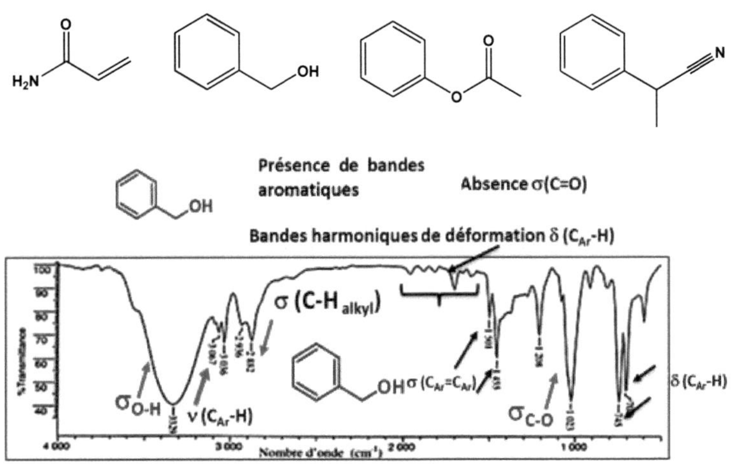

Présence d'une bande σ(C=C) à 1617 cm-1

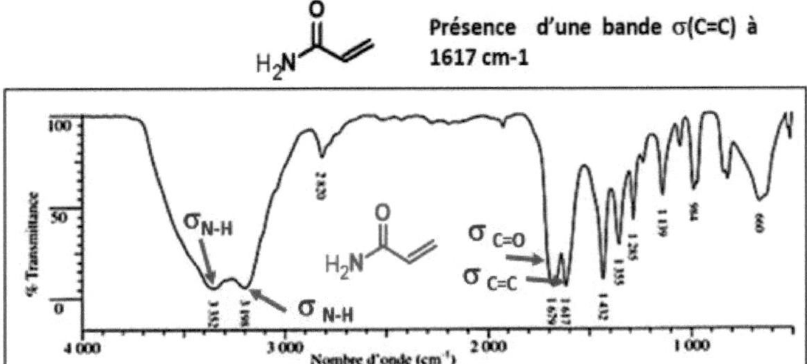

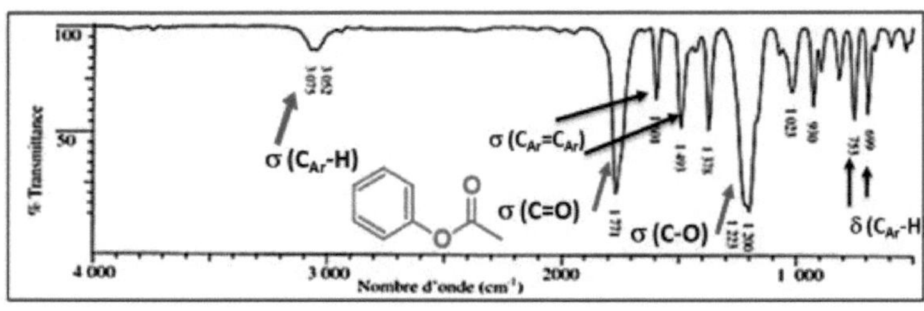

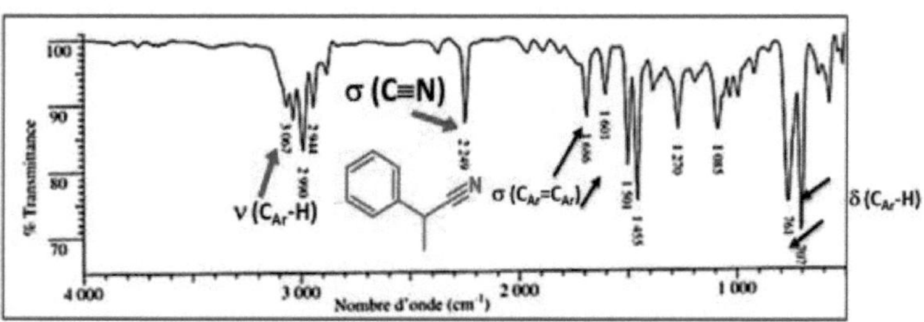

References

1. W.S. Lau, *Infrared characterization for microelectronics*, World Scientific, 1999.

2. P. Hamm, M. H. Lim, R. M. Hochstrasser, "Structure of the amide I band of peptides measured by femtosecond nonlinear-infrared spectroscopy", *J. Phys. Chem. B*, vol. 102, 1998, p. 6123 (DOI 10.1021/jp9813286).

3. S. Mukamel, "Multidimensional Fentosecond Correlation Spectroscopies of Electronic and Vibrational Excitations", *Annual Review of Physics and Chemistry*, vol. 51, 2000, p. 691 (DOI 10.1146/annurev.physchem.51.1.691)

4. N. Demirdöven, C. M. Cheatum, H. S. Chung, M. Khalil, J. Knoester, A. Tokmakoff, "Two-dimensional infrared spectroscopy of antiparallel beta-sheet secondary structure", *Journal of the American Chemical Society*, vol. 126, 2004, p. 7981 (DOI 10.1021/ja049811j)

5. Bee, K. B., Grabska, J., & Huck, C. W. (2020). Physical principles of infrared spectroscopy. Comprehensive Analytical Chemistry. doi: 10.1016/bs.coac. 2020.08.001.

Blinder, S. M. (2004). Molecular Spectroscopy. Introduction to Quantum Mechanics, 217-241. doi:10.1016/b978-0-12-106051-0.50020-x

6. Dagdigian, P. J. (2014). Fundamentals of optical spectroscopy. Laser Spectroscopy for Sensing, 3-33. doi:10.1533/9780857098733.1.3

7. Dutta, A. (2017). Fourier Transform Infrared Spectroscopy. Spectroscopic Methods for Nanomaterials Characterization, 73-93. doi:10.1016/b978-0-323-46140- 5.00004-2

8. Harris, D. C. (2011). Quantitative chemical analysis. New Y o r k , NY: W.H. Freeman and Co.

9. Owen, A., J. UV-Visible Spectroscopy: Uses of Derivative Spectroscopy. Agilent Technologies.

10. User's manual UviLine 9100 - 9400, Ref. 0M8626. SECOMAM, a NOVA ANALYTICS Company. ALES FRANCE.

11. Workbook. Fundamentals of UV-visible spectroscopy. Agilent Technologies 2000.

12. Rouessac, F. Rouessac. A. Cruché, D. (2004). Chemical Analysis - 6th edition: Modern methods and instrumental techniques. Édition DUNOD.

13. Rouessac, F. Rouessac. A. Cruché, D. Martel, A. (2019). Chemical analysis - 9th edition: Methods and instrumental techniques. Édition DUNOD

14. Rouessac, F. Rouessac. A. (2021). L'essentiel de techniques instrumentales d'analyse chimique. Édition DUNOD

15. Bernard, A.-S. Clède, S. Émond, M. Monin-Soyer, H. Quérard, J. (2012). Techniques expérimentales en CHIMIE. Édition DUNOD

16. Bernard, A.-S. Clède, S. Émond, M. Monin-Soyer, H. Quérard, J. (2018). Techniques expérimentales en chimie - Classes prépas et concours - 3e édition. Édition DUNOD

17. Bernard, A.-S. Clède, S. Émond, M. Monin-Soyer, H. Quérard, J. (2023). Techniques expérimentales en chimie - Classes prépas et concours - 4e édition. Édition DUNOD

18. House, J. E. (2018). Molecular Rotation and Spectroscopy. Fundamentals of Quantum Mechanics, 137-158. doi:10.1016/b978-0-12-809242-2.00007-3

19. House, J. E. (2018). Molecular Spectroscopy. Fundamentals of Quantum Mecharucs, 271-296. doi:10.1016/b978-0-12-809242-2.00011-5

20. Katle, B. P. (2020). Infrared (IR) spectroscopy. Chemical Analysis and Material Characterization by Spectrophotometry,199-243. doi:10.1016/b978-0-12-814866- 2.00007-5

21. Kenouche S. (2016-2017), DEPARTMENT OF MATERIAL SCIENCES, Université M. Khider de Biskra, Optical spectrometry: theory and experiment. Course syllabus

22. CHIGR M., (2019-2020), Ecole Supérieure de Technologie Fkih Ben Salah Université Sultan Moulay Slimane, Spectroscopic methods. Course handout

23. M.Hamon, F.Pellerin, M.Guernet, G.Mahuzier. (1998) Chimie analytique Tome 3 Méthodes spectrales et analyse organique. Masson.

24. Ni, L., Geng, X., Li, S., Ning, H., Gao, Y., & Guan, Y. (2019). A flame photometric detector with a silicon photodiode assembly for sulfur detection. Talanta, 120283. doi:10.1016/j.talanta.2019.120283

25. Ouellette, R. J., & Rawn, J. D. (2018). UV-Visible and Infrared Spectroscopy. Organic Chemistry, 409-425. doi:10.1016/b978-0-l2-812838-1.50014-1

26. Poole, C. F. (2016). Detectors. Reference Module in Chemistry, Molecular Sciences and Chemical Engineering. doi:10.1016/b978-0-12-409547-2.1 1719-6

27. Smith, B. C. (2002). Fundamentals of Molecular Absorption Spectroscopy. Quantitative Spectroscopy: Theory and Practice, 1-41. doi:10.1016/b978- 012650358-6/50002-6

28. Stephanos, J. J., & Addison, A. W. (2017). Vibrational rotational spectroscopy. Electrons, Atoms, and Molecules in Inorganic Chemistry, 505-584. doi:10.1016/b978-0-12-811048-5.00009-2

29. GUENNOUN, L. EL HAJJI, A. (2020-2021), FACULTE DES SCIENCES, UNIVERSITE MOHAMMED V, Techniques Spectroscopiques d'analyse (SMC5). Course handout

30. Dr. BELAID Kumar, Faculty of Technology, Université Djillali LIABES Sidi-Bel-Abbes, (2020-2021), Analytical techniques (LGP5). Course handout

31. Silverstein, R. M. Webster, F. X. (1997). Spectrometric Identification of Organic Compounds, 6th Edition Wiley,

32. Silverstein, R. M. Webster, F. X. Kiemle, D. J. (2005). Spectrometric Identification of Organic Compounds, 2eme Edition de Boeck,

33. Silverstein, R. M. Webster, F. X. Kiemle, D. J. (2005). Spectrometric Identification of Organic Compounds, 7th Edition Wiley,

34. Silverstein, R. M. Webster, F. X. Kiemle, D. J. Bryce, D. L. (2014). Spectrometric Identification of Organic Compounds, 8th Wiley Edition,

35. Barnes, J.D. Denney, R.C. Mendham, J. Thomas, M.J.K. 6th Edition (English). Translator: Mottet, M. Toullec. J. (2005). Vogel's Quantitative Chemical Analysis, 1ere Edition de Boeck

36. Harvey, D. (2000), Modern Analytical Chemistry chemistry, - 1st ed, The McGraw-Hill Companies

Printed by Books on Demand GmbH, Norderstedt / Germany